EARTH FOR SALE

ALSO BY MAUDE BARLOW

Blue Gold:
The Battle Against Corporate Theft of the World's Water

Blue Covenant:
The Global Water Crisis and the Coming Battle for the Right to Water

Blue Future:
Protecting Water for People and the Planet Forever

Boiling Point:
Government Neglect, Corporate Abuse, and Canada's Water Crisis

Whose Water Is It, Anyway?
Taking Water Protection into Public Hands

Still Hopeful:
Lessons from a Lifetime of Activism

MAUDE BARLOW

EARTH FOR SALE

THE FIGHT TO STOP THE LAST PLUNDER OF THE PLANET

ECW

Published by ECW Press
665 Gerrard Street East
Toronto, Ontario, Canada M4M 1Y2
416-694-3348 / info@ecwpress.com

Editor for the Press: Susan Renouf
Copy editor: Jennifer Knoch
Cover design: Jessica Albert
Cover image: Photo of Earth taken by the Suomi NPP Satellite in 2012, courtesy of NASA

LIBRARY AND ARCHIVES CANADA CATALOGUING IN PUBLICATION

Title: Earth for sale : the fight to stop the last plunder of the commons / Maude Barlow.

Names: Barlow, Maude, author

Identifiers: Canadiana (print) 2026011989X | Canadiana (ebook) 20260119962

ISBN 978-1-77041-866-0 (softcover)
ISBN 978-1-77852-572-8 (PDF)
ISBN 978-1-77852-571-1 (ePub)

Subjects: LCSH: Environmental economics. | LCSH: Environmental protection—Economic aspects. | LCSH: Commodification.

Classification: LCC HC79.E5 B37 2026 | DDC 333.7—dc23

This book is funded in part by the Government of Canada. *Ce livre est financé en partie par le gouvernement du Canada.* We also acknowledge the support of the Government of Ontario through the Ontario Book Publishing Tax Credit, and through Ontario Creates.

PRINTED AND BOUND IN CANADA

PRINTING: MARQUIS 5 4 3 2 1

To all the waters of the Earth

CONTENTS

Climate breakdown is the result of a global failure to address the power of private capital.

— GEORGE MONBIOT, *GUARDIAN* COLUMNIST

The Earth is a living, breathing entity, just the same as our bodies are. Our survival utterly depends on living in nature, not apart from it. In addressing climate change, we need to move away from focusing solely on the language of economics, which further adds to the destruction of our atmosphere, our land, our waters and wildlife, and we need to emphasize and consider the impact on human life and rights as well. Climate change is very much about the moral and ethical imperative.

— SHEILA WATT-CLOUTIER,

INUK ENVIRONMENTALIST AND ACTIVIST

INTRODUCTION

This book is about a new iteration of an old problem: the destruction of our planet — its rivers, forests, soil and air — by those seeking to profit from its exploitation. For too long, too many of us have seen nature as a vast resource for our pleasure and convenience and as the fuel for unlimited economic growth. For too long, we have watched as transnational corporations have used nature as their personal property and path to wealth.

In some important ways, we are collectively waking up to the fact that there is no life worth living on a dying planet. Powerful movements have grown to protect the environment. Environmental rights and the human right to a healthy environment have been enshrined in international law. Under the sponsorship of the United Nations, international gatherings on the climate, water and biodiversity crises are held regularly and the world has come together to endorse the UN Sustainable Development Goals that link human progress and equality to a sustainable environment. Indigenous rights to protect the resources in their territory have been enshrined in the UN Declaration on the Rights of Indigenous Peoples.

There has also been an awakening at the highest levels of leadership about the urgent need to expand the climate change fight by ramping up the protection and restoration of biodiversity. While it is clearly important to continue reducing greenhouse gas emissions and transitioning to climate-friendly energy alternatives, it is becoming increasingly evident that damaged watersheds, wetlands, forests and soil are major contributors to the climate crisis and their care and restoration will create a new path to climate stability. Healthy forests and soils are carbon sinks. Restored watersheds and wetlands green deserts, bring back the rain and cool local hydrologic cycles.

The good news is that there is a tsunami of international gatherings, pledges, programs and commitments being made to promote biodiversity as a way to address the climate crisis; trillions of dollars have been earmarked for this "new" frontier.

The bad news — and this is the new iteration of the old plunder — is that many in the private sector are looking to access these funds. Transnational corporations, global equity funds, asset managers, large agribusiness, energy and chemical companies, as well as private water utilities and bottled-water companies, are keen to become players in the new transition to what is called "nature-based solutions." The same corporations and industries that brought us to the ecological brink have jumped on the "nature" bandwagon and are appropriating the language of environmental groups and the UN to show their commitment to the need to protect biodiversity.

Their answer to the climate crisis: bring nature into the market, put a price on it, and let the market — not governments — guide the process. Where governments use regulatory measures to protect nature, the market treats it as an asset to be bought, traded and sold. This latest threat to the planet and human survival is the financialization of nature: carbon trading, water pollution trading, biodiversity credits, wildlife conservation bonds, nature bonds, plastics offsets, nature-based solutions, green growth, natural capital, water futures, ecosystem services. This is the new language of the multitude of private interests keen to profit from the growing global commitment to protecting nature.

British economist Sir Dieter Helm approvingly explains in the 2014 report *The State of Natural Capital*: "The environment is part of the economy and needs to be properly integrated into it so that growth opportunities will not be missed."[1]

The financialization of nature is a dangerous development that surprisingly few know about, including many in the environmental and conservation movements. This book sounds the alarm on this insidious new form of corporate plunder. We must not allow the movement to protect nature and biodiversity to be taken out of the hands of elected governments and communities and handed over to corporations and global investment funds, who would bring nature into the market where profit is, and always has been, the driving motive.

CHAPTER ONE

THE ENCLOSURE OF THE COMMONS

The land knows you, even when you are lost.

— ROBIN WALL KIMMERER,
POTAWATOMI BOTANIST AND WRITER

In a moving and groundbreaking June 2024 ceremony in Rarotonga, Cook Islands, Indigenous Pacific Island leaders signed a new oceans treaty officially recognizing whales and dolphins as legal persons. Whales now have inherent rights, said the declaration, such as freedom of movement, the right to a healthy environment and the right to thrive alongside humanity. With this treaty, say its founders, the giant of the seas will be protected in South Pacific waters by legislation and its habitat secured. "The sound of our ancestor's song has grown weaker, and her habitat is under threat. . . . He Whakaputanga Moana [Declaration for the Ocean] is not merely words on paper. It's a Hinemoana Halo, a woven cloak of protection for our taonga — our treasures — the magnificent whales," said Aotearoa (New Zealand) King Tūheitia. While not yet binding, the treaty has allowed the Māori and other Indigenous communities to start talks with governments.[2]

The treaty has friends in high places. In November 2024, Great Britain's King Charles welcomed Mere Takoko, Māori leader and chief executive of the Pacific Whale Fund, to St. James's Palace to take part in the creation of a new Bioeconomy Alliance. The ocean treaty is part of the fast-growing movement around the world to protect nature by giving it legal rights and having guardians — often Indigenous people — as protectors. It is being heard in the halls of power.

Alongside the Indigenous leaders who signed the treaty in Rarotonga that day were marine scientist Dr. Carlos Duarte and economist Dr. Ralph Chami. Duarte, currently teaching at King Abdullah University of Science and Technology in Saudi Arabia, is well-known in international circles and conferences for his work on the restoration of mangroves and his promotion of "blue carbon," a plan to use the oceans and marine ecosystems to sequester carbon from the atmosphere. While Duarte is touted by many as a leading climate scientist, others are questioning his acceptance of funds from Saudi Arabia — a petrostate — to back his research, as well as the dangers of transferring the carbon crisis from the air to the oceans. Especially concerning is his support of carbon offsets to fund his ocean protection plan.

Dr. Ralph Chami is currently on leave from the International Monetary Fund (IMF), where he has served as a senior economist for almost 30 years. He is known for his promotion of valuing wild animals such as elephants and whales as carbon sinks, arguing that if their role in sequestering carbon was more widely known, they would be more likely to be protected. Like Duarte, Chami promotes carbon offset trading, whereby polluters can continue to pollute as long as they "offset" their damage by supporting projects elsewhere.

Chami founded Blue Green Future, a group of "innovators and thought leaders for natural capital finance and a nature-based economy in order to translate ecosystem services into monetary values." On the potential to harness wild animals into the market, Chami says, "Incorporating the carbon services of wild animals into financial markets has the potential to benefit both climate and conservation through the development of carbon offsets that are equitable and nature positive."[3] He calculates an elephant is worth US$1.75 million and a whale is worth US$2 million.

Why were a leading scientist and a leading economist who espouse market solutions to the climate crisis in attendance at a Māori ceremony in the Cook Islands dedicated to enlarging the rights of whales?

Privatizing nature to save capitalism

The Earth is in peril. It is well-documented that our planetary crisis includes climate chaos due to runaway carbon emissions, rapidly dwindling clean water supplies, warming of the oceans, mass extinction of species and historic biodiversity loss. What is desperately needed is a commitment to stop the destruction of nature. This would require governments and international institutions to regulate its protection. Also desperately needed is a radical rethinking of our economic values and systems, most especially the myth that unlimited growth is somehow sustainable.

While some very important initiatives to protect nature, restore biodiversity, transition to clean energy and protect the human right to a healthy environment have come from the United Nations and many nation states, a competing narrative has grown up in the last decade that is threatening to replace government responsibility and cooperative regulation. It is important to remember that the UN is a contested venue where many conflicting ideologies and interests compete to be heard. While the UN has a rich history with the creation of the Sustainable Development Goals, powerful corporate interests and neoliberal governments that promote them have a very strong presence and growing influence there.

The message from some high-level committees and agencies of the UN, as well as the World Bank, the Organisation for Economic Co-operation (OECD), the European Commission and some wealthy governments, is that we can keep on with business as usual as long as we bring nature into the market where its protection can be overseen by experts in the workings of the market. While their motives might truly be to protect the elephants and the whales, Duarte and Chami are promoting a market-based solution to the environmental crisis that threatens to be just as dangerous to the planet as the plunderers of old.

This solution — dubbed false by many human rights and environmental organizations — is variably called the financialization of nature, green growth, ecosystem services, green economy, natural capital, nature-based solutions, green capitalism and conservation finance.

In her groundbreaking 2022 book, *The Value of a Whale*, Adrienne Buller, senior fellow with the British think tank Common Wealth, writes that green capitalism is an effort to preserve existing capitalist systems by addressing environmental catastrophe through new paths to accumulation, while minimizing disruption to our current economic system and ways of life. She calls this a new iteration of the "Wall Street Consensus," a coordinated effort to sell new financial opportunities to the market and "escort private capital into new desired spaces." In practice, this means financial incentives in lieu of firm regulation, and the utilization of private finance rather than direct public or governmental investment. She writes:

> It seems that in its relentless pursuit of new opportunities for investment and fixation on risk, finance has finally turned to a new frontier: nature. Betting that ecosystems could be to the 2020s what subprime mortgages were to the noughties, financial institutions have been laying the groundwork for the emergence of "nature as asset class" — that is, the transformation of the natural world into a new suite of tradable assets, and hoping that through securitization (taking a group of illiquid assets, like mortgages, and bundling them together into a security) the risks of its collapse can be mitigated.

> The language of "natural capital" and "ecosystem services" now tidily segments the complexity of nature and biodiversity into discrete, measurable and — critically — costed units, a clever sleight of hand that opens the door to the commodification of the natural world while passing as an innovative mechanism for its protection. Ecosystems, its advocates argue, are being destroyed because they currently have no economic value; the right price tag could be their saviour . . . By establishing rights to pollute the atmosphere that can be bought, sold and converted into novel financial assets, new pathways for profit and speculation are created, while the need for states to impose meaningful regulation can be met in a cosmetic sense without the risk of alienating the corporate and capitalist base to which they are in various ways beholden.[4]

Green capitalism is not to be confused with outright denial that there is a serious ecological crisis. In the recent past, major energy companies have lobbied hard to convince legislators and the public that reports of a climate crisis are greatly exaggerated and that the world still needs fossil fuels. But more recently, Big Oil has changed its public face on climate change. With brochures and websites teeming with photos of soaring eagles, pods of healthy whales and old-growth forests, the companies have publicly embraced nature and committed themselves to its protection. Buller argues that this new corporate face is more dangerous even than outright denial as it not only allows the industry to continue to exist, grow and pollute, it gives it great say in the fashioning of market-friendly climate solutions.

In a 2018 opinion piece for *The Guardian*, George Monbiot called the financialization of nature a "magic powder" to allow governments to avoid the hard choices, laws and policies they should be making to deal with the many environmental crises we face by replacing political decisions with economic calculations. He writes that it is deluded to expect we can defend the living world through the mindset that is destroying it.

> The notions that nature exists to serve us; that its value consists of the instrumental benefits we can extract; that this value can be measured in cash terms; and that what can't be measured does not matter, have proved lethal to the rest of life on Earth. The way we name things and think about them — in other words, the mental frames we use — helps determine the way we treat them. . . . The natural capital agenda is the definitive expression of our disengagement from the living world. First we lose our wildlife and natural wonders. Then we lose our connections with what remains of life on Earth. Then we lose the words that described what we once knew. Then we call it capital and give it a price. This approach is morally wrong, intellectually vacuous, emotionally alienating and self-defeating.[5]

Antonio Tricarico, the Italian energy researcher and writer who first coined the term "financialization of nature," warned that this trend could put the management of the commons into the hands of financial markets for years to come. "This approach is a long-term project that aims to lock natural resources management into the future structure of capital markets in a way that will dramatically reduce the possibilities to reclaim the commons and their collective management by affected communities. This systemic financial enclosure of the commons, coupled with existing trade and investment liberalization agreements, would produce a long-lasting legal enclosure that drastically shrinks the political space for social movements to reclaim the commons as the basis of their livelihoods," he told the Alternative Water Forum in 2012.[6]

Plundering the commons

An April 2025 study by a team from Stanford University and Dartmouth College reported in the journal *Nature*, found that the world's biggest 111 companies — over half being fossil fuel providers — caused US$28 trillion in climate damage between 1991 and 2020. For comparison, said the researchers, this is a shade less than the sum of all goods and services produced in the United States in 2024. This does not include the costs of extreme weather caused by the climate crisis, such as droughts, floods and hurricanes.[7]

A large part of our environmental crisis is the result of investment by corporations and governments in energy and mining extraction, overfishing, forest felling and soil exploitation. For centuries, we humans have ruthlessly used the planet's natural wealth as a source of unlimited growth, fuelling an economy based on the subsequent demand for ever-more resources. In doing so, we have relentlessly diminished the space for local communities to defend their own future and protect the resources in their care.

University of British Columbia professor emeritus William Rees studies the impact on the natural world of the growing footprint of humans. He says that our innate desire and capacity to expand using whatever resources we need has put that world in jeopardy. While humans comprise only 0.01% of the Earth's biomass, our expansion has eliminated the biomass of 83% of wild animals and half of all natural plants. "Indeed," he wrote, "*H. sapiens* may well be the most voraciously successful carnivorous and herbivorous vertebrate ever to walk the Earth — but at the expense of thousands of other species."[8]

Rees is only one of many scientists and scholars who have warned us for years that we are at an ecological tipping point that requires urgent action. And yet, we continue to invest in the very industries and technologies that further destroy the planet and our health. A December 2023 report, *State of Finance for Nature* by the UN Environment Programme, found that close to US$7 trillion is invested each year from both public and private sector sources in activities that have a direct negative impact on nature — *30 times more* money than flows to investments that protect nature and biodiversity.[9] How has this come to be?

The current push to capture and commodify nature is, of course, not the first manifestation of this trend. The plunder of the commons goes back centuries.

What is the commons? The late Canadian environmentalist Richard Bocking said that the commons are those things to which we have rights simply by being a member of the human family. "The air we breathe, the fresh water we drink, the seas, forests and mountains, the genetic heritage through which all life is transmitted, the diversity of life itself." "Commons" is synonymous with community, cooperation and respect for the rights and preferences of others, he wrote.[10]

American journalist Jonathan Rowe explained in *Utne Reader* that the commons is the realm that lies outside both the economic market and the institutional state (although the state has an important role in protecting the commons) that all of us typically use without toll or price. "The atmosphere and oceans, languages and culture, the stores of human knowledge and wisdom, the informal support systems of community, the peace and quiet we crave, the genetic building blocks of life — these are all aspects of the commons."[11]

Two important Western historical markers of the commons were the 1215 Magna Carta, which limited the power of the British monarch, and the 1217 Charter of the Forest, which ensured the rights of people to access the royal forests and woodlands for food foraging, fuel and heating. Woodland management was organized under customary agreements, and seeds were shared for the community. Key to the success of commons' communities was living in — not apart from — nature and the understanding that, to protect the commons for people, it was necessary to protect the land and water that sustained it. Nobel Prize winner Elinor Ostrom studied hundreds of commons-practising communities around the world and found that commons stewardship based on participatory decision-making and clear rules governing the use of common resources is paramount in protecting the environment. While Garrett Hardin's infamous 1968 *Science* "tragedy of the commons" article claimed that commons overexploitation is inevitable, subsequent research, including Ostrom's, has shown that, while not all commons management is without flaws, well-managed commons stewardship is the best for people and the planet.

Many Indigenous societies to this day cannot conceive of denying members of the community basic access to food, land and livelihood. There are still large parts of the world where people live in commons-structured communities sharing land and resources, and where the rules of the global market do not apply. That is, until they are invaded.

Colonial resource grab

The often violent colonization of the Americas at the end of the 15th century, and the subsequent spread of disease, disrupted the agricultural practices of the local Indigenous populations and killed so many people it disturbed Earth's climate, creating a mini ice age, a 2019 study published in *Quaternary Science Review* by scientists at University College in London has found. The disruption to the commons that followed Christopher Columbus's initial journey led to the death of 56 million Indigenous people by 1600 and the abandonment of vast tracts of community-farmed land.[12]

The following centuries saw ruthless colonialist plundering of the Americas. Foreign states and later corporations voraciously extracted natural resources, disrupting local, Indigenous, peasant and small farmers' communities and creating horrific human rights violations. In Uruguayan writer Eduardo Galeano's 1971 book, *Open Veins of Latin America: Five Centuries of the Pillage of a Continent*, he detailed the rapacious plunder of Latin American gold, silver, tin, copper, oil, nitrates and rubber through indentured slavery, extractivism and colonization. He would later write about the recolonization of his people through forced debt, unfair trade and corporate impunity.

The plunder continues to this day. A 2024 report by Brazilian human rights group Conectas found that modern-day slavery still exists in the coffee supply chains of foreign companies, including Nestlé, in that country, where thousands of workers toil in dangerous conditions "akin to slavery." Starbucks, the world's leading coffee brand, is currently being sued in Washington for "severe human rights and labour abuses" in its Latin American operations by the National Consumers League, a consumer advocacy organization.[13]

The Center for Human Rights of the American Bar Association reports that illegal mining in Latin America poses severe threats to international law, human rights and the environment. In the Amazon rainforest, illicit mining of gold drives deforestation, habitat destruction and biodiversity loss. Mercury contamination has destroyed major rivers and caused severe health problems for local communities in many countries. In Venezuela, armed groups control mines with brutal force, often resorting to extreme violence, including torture and murder, to enforce compliance. Children are working in deep pits, handling toxic substances without protective gear. Environmental defenders in Latin America face the gravest risks and are frequent targets of violence and murder.[14]

In its first six decades of existence, the US created a trail of treaties that led to a massive transfer of Indigenous resource wealth, writes Michael John Witgen, a Lake Superior Ojibwe member and history professor with Columbia University, in his 2022 book, *Seeing Red: Indigenous Land, American Expansion, and the Political Economy of Plunder in North America*. The US coerced Indigenous Peoples to surrender vast territories, including forests, waterways and land that would create the conditions for explosive economic growth for the country. "The total compensation to Indigenous people amounted to pennies per acre for Native territories," writes *People's World* in a review of the book. "The treaty procedure was propelled by threats of war, forced removal, and simply taking the land without compensation."[15]

Throughout the early 20th century, the federal and provincial governments in Canada similarly systematically displaced the first peoples of the land and claimed the immense wealth of the water, forests, fish, minerals and land in their territories. The Assembly of First Nations reminds us that the Indian Act of 1876 officially banned First Nations people from land ownership and allowed companies to extract resources on reserve land for nominal or no compensation at all. Loggers pillaged the old-growth forests in British Columbia so ruthlessly, there are many outstanding claims for reparations to this day.

In northwestern Ontario, the Ojibwe of Grassy Narrows were displaced from their 142,000 square kilometres of land and moved to a reserve of just over 41 square kilometres, writes Mount Allison professor David Thomas in his 2022 book, *Capitalism & Dispossession: Corporate Canada at Home & Abroad*. This cleared the way for large-scale mining and forestry and left the people to subsist on fish and game even as the Dryden paper mill dumped nearly 10 tonnes (9,000 kilograms) of mercury into the English-Wabigoon River systems, causing illness and death from mercury poisoning to continue.

Before the Industrial Revolution, many rural communities in Europe were organized as commons, producing food largely in harmony with nature. The first great English enclosure of the commons took place in 1740, when common grazing land was seized and privatized. Indian physicist and environmentalist Vandana Shiva writes that the privatization of the commons supplied a steady source of raw materials that fuelled the Industrial Revolution. Today, half of England's land is owned by less than 1% of its population.[16]

According to Utsa Patnaik, professor emeritus at New Delhi's Jawaharlal Nehru University, the British drained approximately

US$45 trillion from India between 1766 and 1938, a sum 17 times larger than the current GDP of the UK.[17] While Patnaik's report has generated much discussion, including questions about how to measure the economy of two centuries ago using today's model, there can be no doubt that the resources of India were taken for trade by powerful forces outside the country. Much of the country was organized around rural communities structured as commons, which have been relentlessly challenged by the economic invaders. In 1865, a law in India that protected forests as commons was lifted, paving the way for the commercial exploitation of both land and forests.

In Australia, the concept of terra nullius (literally meaning "empty land") was used to justify the appropriation of land and its natural resources by declaring the entire continent of Australia uninhabited, Shiva wrote for the Third World Network. "This declaration enabled the colonizers to privatize the commons relatively easily, because as far as they were concerned, there were no commons in the first place."[18]

In Africa, European colonizers claimed land as a reward of conquest, replacing the Indigenous way of food production that operated within a tradition of respect for nature, with one that mercilessly cleared land, forests and wetlands for industrial farming, write community water organizers Minni Jain and Philip Franses in their 2025 book, *The Language of Water*. European powers also ruthlessly plundered other African resources. Shell Oil set up operations in Nigeria in 1938 and started pumping oil for export two decades later. The Nigerian National Oil Spill Detection Agency holds Shell responsible for hundreds of oil spills that have destroyed large areas of farmland and killed fish and wildlife. The Centre for Research on Multinational Corporations says that Shell has profited greatly for over half a century, leaving the Niger Delta as "one of the most oil-polluted places on Earth."[19]

The world learned of the extent of human rights abuses connected to Shell in 1995 when the Nigerian government executed anti-Shell activist Ken Saro-Wiwa and eight other Ogoni activists. In December 2024, the Nigerian government gave Shell permission to sell its operations in Nigeria, angering many who say the company must take responsibility for its horrific environmental footprint before it pulls out. Amnesty International says that it is essential that Shell be held accountable for decades of "grievous human rights and environmental abuses."[20]

In the first decade of the 21st century, wealthy countries and large corporations have grabbed so much land to grow food for export, the United Nations introduced a plan to prevent uncontrolled purchase of land by foreign investors in the Global South. However, a 2021 report by the Land Matrix Initiative found that between 2010 and 2020, 7.3 million hectares of land in sub-Saharan Africa alone were leased by foreign investors.[21] The enclosure of this land also gave foreign countries and large corporations control over the water on and under that land, essentially privatizing vast amounts of local water sources. Water was also enclosed and commodified through long-term mining contracts, whereby foreign mining companies gain access to local watersheds to dump their water tailings.

Economic resource grab

The modern enclosure of the commons started with economic globalization, the explosive growth of transnational corporations and the introduction of free trade agreements that limited the ability of governments to regulate and protect their forests, minerals, fisheries and water sources — now often labelled as tradable commodities — ceding their fate to the market.

British political economists Jason Hickel and Dylan Sullivan have observed that after most imperial powers withdrew their flags and armies from their former colonies, the pattern of colonial appropriation remained in place and continued to define the global economy: "Imperialism never ended, it just changed form." The "hidden transfer of wealth" from poor countries to rich countries and corporations took place without the overt violence of colonial occupation, they explained, but through a global market economy that continued to rely on a large net expropriation from the Global South, including tens of billions of tonnes of raw materials and hundreds of billions of hours of human labour per year. In a 2021 *Al Jazeera* article, they maintained that since 1960, rich countries have drained US$152 trillion from the Global South.[22]

An important feature of the global market economy that facilitated this later resource grab was the transfer of authority over natural resources from governments to the private sector through deregulation and privatization. Steven Bernstein, a political scientist at the University of Toronto, argues that a market framework for environmental policy was already formed in the late 1970s and 1980s by experts in the OECD's environment committee, and they brought the liberal market consensus to the Brundtland Commission in the 1980s. Market-based solutions were well-established when the climate issue made its debut at the Rio UN Conference on Environment and Development in 1992. In his book, *The Compromise of Liberal Environmentalism*, Bernstein writes, "By 1992, a shift in norms of environmental governance had occurred, characterized by a general acceptance of liberalization in trade and finance as consistent with, and even necessary for, international environmental protection. These norms also promoted market and other economic mechanisms (tradable pollution permits or the privatization of the commons) over 'command-and-control methods' (standards, bans, quotas, and so on) as the preferred method of environmental management. The concept of sustainable development, while it legitimated this shift in norms, now masks this compromise that characterizes international environmental governance."[23]

The World Bank and the International Monetary Fund, the two institutions established at the end of World War II to rebuild the international economic system, also changed to reflect the new economic order. The London-based Bretton Woods Project reports that the original apolitical mandate of the two institutions was transformed in the late 1970s to align with what would later be called the Washington Consensus to focus on free-market economic policies such as free trade, deregulation, privatization and unlimited growth. The watchdog think tank documents the harm caused by these institutions, including mass evictions and the forced displacement of local communities and marginalized people.[24]

At this time, the World Bank and the International Monetary Fund started to impose free-market structural adjustment programs (SAPs) on countries of the Global South as a condition for receiving loans. The stated aim was to make the recipient countries more "competitive" by devaluing currency, cutting public spending and privatizing and deregulating industry, especially those extracting natural resources. The result was socially, economically and ecologically disastrous. Poor countries racked up huge, ever-growing debts, and some found themselves spending more on debt repayment than on health care or education. The price of debt repayment was to open their labour, markets and natural resources to foreign corporations, who moved in to take over the available riches.

SAPs prohibited poor countries from setting limits on foreign investment, giving corporations of the Global North almost unlimited access to the forests, minerals, land and water of the recipient nations. A summer 2003 publication of the Washington-based Institute for Policy Studies reported that the World Bank had provided hundreds of billions of dollars to subsidize the efforts of global corporations that allowed them to establish control over the natural resources in many countries of the Global South.

In fact, many of the roads, electrical grids and power plants the bank boasted it was building in poor countries were built primarily to serve the foreign resource extraction companies, not local communities. According to the institute, by 2000, the World Bank, through its support of fossil fuel projects in the Global South, became the largest contributor to greenhouse gas emissions in the world.[25]

Corporate power grab

The narrative that a deregulated global market economy was both good and inevitable became entrenched as transnational corporations grew in size, number and power. Transnational corporations (TNCs, also called multinational corporations) are companies that have essentially outgrown their country of origin and operate in several countries, perhaps having their financial head office in one country, their tax haven in another and production in a number of countries of the Global South. According to the United Nations Conference on Trade and Development, TNCs have exploded in the last half century. In 1970, there were around 7,000 TNCs; by 2018, there were over 80,000, and they generated over two-thirds of international trade.[26]

Their wealth is immense. Global Justice Now, a UK-based social justice organization, reported in 2018 that the largest 2,000 TNCs had a combined revenue of US$48 trillion. This gives these corporations a great deal of power in determining the rules of international trade and resource extraction. In fact, 69 of the top 100 economies in the world by revenue are corporations, not countries, according to Oxfam and Global Justice Now. Corporate revenues of Apple, BP and Shell exceed the revenues of Russia, Belgium and Sweden. Exxon's exceed India's. Walmart's revenues exceed the national government revenues of two-thirds of the world's countries. China National Petroleum is richer than South Korea and Mexico.[27]

As they left behind their home base, these companies sought what they call a "level playing field" — uniform rules that make it easy to cross national boundaries without being held up by the "red tape" of different regulations and standards. Toward this end, large corporations were instrumental in the creation of free trade agreements that bind governments to business-friendly rules. Until the last few years, the growth of free trade bilateral and regional agreements has been explosive, mirroring the rapid growth of TNCs.

Prior to the 1980s, there were only a handful of free trade agreements in the world. By 1990, there were 50, and by 2022, there were

355 in force and another 500 either adding new members or negotiating new agreements.[28] The World Trade Organization (WTO) was created in 1995 to oversee the global trading system, founded on the doctrine that the market, not governments, should set the rules of trade.

Transnational corporations wanted (and got) several key things from free trade agreements: to be able to move production to low-wage countries without censure; to remove export controls on resources such as trees, minerals, fish, food, seeds and energy to supply their global manufacturing chains and to challenge rules protecting the environment and workers' rights as "unfair trade practices."

For decades, Indigenous and small landholders in the Global South have reported on the negative impact of the free trade agenda on their lives. The US-based Indigenous Environmental Network has documented how the WTO has forced poor countries and rural communities to open their borders to transnational agricultural companies for commercial food production, displacing traditional plant varieties with genetically altered species and ultraprocessed foods. Of huge concern has been the theft of local, community and Indigenous knowledge through the intellectual property provisions of trade deals.

Seed patenting by transnational corporations poses a direct threat to biodiversity, says Vandana Shiva in *The Enclosure and Recovery of the Commons*, transferring traditional knowledge held by the community to private corporations in the form of patents. Thus, "the corporation is now treated as the only form of association with legal personality," she wrote.[29] In fact, corporations in the United States have legally recognized personhood rights, including property rights, that can be used to challenge laws regulating their behaviour.

Parallel to the growth of free trade agreements, bilateral investment treaties suddenly proliferated, giving private investors and foreign corporations the right to challenge the environmental, health and labour regulations and standards of the signatory countries. With investor–state dispute settlement (ISDS), foreign investors can sue countries in secretive tribunals when they believe their profits have been harmed by public policies.

There are over 3,000 bilateral investment treaties in the world, adding a powerful mechanism for the private sector to threaten governments. The World Bank's International Centre for Settlement of Investment Disputes has heard over 1,400 cases, the great majority laid against poor countries in Latin America, the Caribbean, Africa and Central Asia, and ruled in favour of the private investors. Compensation settlements have expanded dramatically, and it is not unusual for a poor country to pay out compensation in the billions. The Canadian International Institute for Sustainable Development says the rise in investor–state challenges has deterred many governments from adopting or enforcing environmental regulations.

Former Irish president Mary Robinson and former UN secretary-general Ban Ki-moon warned in a 2024 article in *The Guardian* that fossil fuel companies are using ISDS to force governments to compensate them for lost profits when the governments introduce measures to transition to a low-carbon economy. They expressed their outrage that investment treaties give foreign energy corporations the right to sue for compensation in the billions if governments start to fulfill their climate commitments under the Paris Agreement, adding that such corporate power is "destroying the world's ability to counter harmful activities" of the energy industry.[30]

"We're left with two regimes that conflict: the Paris Agreement requires (broadly) that fossil fuels be phased out, and the ISDS regime provides guarantees for investors that protect their investment — even if it is a fossil fuel investment. That's the conflict — it's as simple as that," British barrister and arbitrator Toby Landau told *The Guardian*. Rob Davies, former minister of trade and industry in South Africa, concurred: "Companies have got the right to challenge any policy . . . that will impact their expectation of profitability in the future, no matter what the regulation is, no matter what its motivation is, no matter how well designed it is or anything." He believes more recently fossil fuel companies are using the ISDS provisions to thwart regulations on green transition. "It has a chilling effect, particularly on developing countries."[31]

It is not just public policies of the Global South that have been challenged. Under the North American Free Trade Agreement, Canada has paid hundreds of millions of dollars to American corporations for environmental policies they claimed were barriers to trade. But Canada is not just a victim of ISDS: in 2022, the Canadian Centre for Policy Alternatives reported that Canadian mining and energy companies, among other investors, had initiated at least 56 ISDS claims and were awarded US$10.2 billion in compensation to date. "These figures represent enormous costs to the countries involved, many of which are very poor . . . The 'regulatory chill' created by these cases poses a direct and immediate threat to sustainable resource development and strong environmental policies," says the report.[32]

A March 2025 investigation by *The Guardian* reported that, since 1998, a stunning US$120 billion of public money has been awarded through ISDS courts to corporations, but it cautioned that the true figure is likely much higher as companies often don't disclose the size of payments they receive. Not surprisingly, fossil fuel companies have been the biggest beneficiaries to date as they challenge laws intended to fight climate change.

These corporate challenges to government rules have become an increasingly popular investment class for hedge funds and asset managers who back them financially in exchange for a share of the final award. Half of all third-party-funded cases were launched by investors from the US, Britain and Canada, and more than three-quarters of cases were against countries of the Global South. The Transnational Institute (TNI) calls this the "financialization of investment law," whereby hedge funds and litigation financiers fuel costly legal battles against states for profit. In claims trading, financial firms buy or broker claims and enforce or resell them on the secondary market. TNI says that the global litigation finance market is estimated to be almost US$90 billion.[33]

A 2025 article in *The Guardian* quotes critics who say that ISDS cases have become big business for asset managers and the extent of the profits is "massive." Third-party funders seek out and invest in cases as a business opportunity, which incentivizes more arbitration, regardless of the broader public interest or the legitimacy of the claims. Philippe Sands, arbitrator and professor at University College London, said, "The arrangements may be said to assist in delivering a flowering of sorts: more fees for the lawyers, more cases for the arbitrators and a deeper, wider and more lucrative trough into which all of our snouts may be dipped."[34]

In "How to Buy a Piece of a Lawsuit and Impoverish a Country," *Inside Climate News* reports how Energy Transition Minerals, an Australian mining company formerly known as Greenland Minerals, is suing Greenland, a semi-autonomous nation of mostly Inuit people, using funds from litigation financier Burford Capital. In 2022, Greenland enacted restrictions on uranium mining after learning how it had "wrought devastation" on Indigenous communities around the world. Energy Transition Minerals is seeking US$11.5 billion in damages, a figure roughly six times greater than the Greenland government spends in a year. The article describes the lead lawyer for the government walking into the first meeting with his three-person team to face 12 corporate lawyers seated over two rows. With possible huge returns on this "investment," Burford, the world's largest publicly listed litigation finance firm, spared no expense. The company's annual performance return is roughly double that of S&P 500 stock index.[35]

A changing corporate climate

There has been a strong backlash to these corporate power and resource grabs. Human rights and environmental justice organizations have fought corporate power and the free trade and investment agreements that underpin their authority. The United Nations Human Rights Council is moving ahead with an international, legally binding instrument to regulate the activities of transnational corporations, although it is very slow in coming. The World Trade Organization has lost a great deal of its power, and many countries are looking to models of self-reliance and security in the face of growing international unrest, climate change and the rise of right-wing governments. All of this is affecting global trade and the bottom line of the corporations that fuel it.

The World Bank reported that in 2024, global trade nearly flatlined, taking a toll on growth everywhere, and 2023 saw the slowest international trade pace in 50 years outside global recessions: "international trade — a key engine of global prosperity since the fall of the Berlin Wall — has ground nearly to a halt and is set to remain anemic in the coming years." In the first half of the 2020s, an average of just five agreements were signed each year — less than half the rate of the 2000s. And the appetite for trade restrictions "seems insatiable," the bank wrote.[36] "Free trade is dead," announced publications as varied as *The Guardian*, *The Atlantic*, and *Foreign Policy* — and that was before President Trump launched his tariff wars, dramatically slowing global trade.

The World Bank blamed "misguided populism" for the new protectionism, and there is some truth in this explanation. Right-wing anti-globalism has grown in recent years, fuelled by conspiracy theories such as the "Great Reset" — the belief that the World Economic Forum planned to use the coronavirus pandemic to enact a one-world government that would force vaccines on everyone and destroy private property rights. (This conspiracy theory is amusing to many left-wing critics of the World Economic Forum, who have long asserted it is a forum to promote corporate capitalism, privatization and a global economy based on a free market.)

President Trump has stoked right-wing populism and heralded the return of more nationalist economic sentiment in other countries with his fervent "America-First" policies. In an attempt to protect themselves from his threats of tariffs and other punishments, some countries have introduced measures enabling them to provide their own populations with food, medicines and basic necessities. This trend started well before Trump's second term, however, when it became clear during the coronavirus pandemic that globalized markets and supply chains miserably failed to provide essential medical supplies, with devastating consequences for many poor countries. Even Doug Ford, Conservative premier of Ontario, has said that his reason for investing in vaccine production in his province was the need to become "self-sufficient."

But much of the backlash against free trade and corporate power has come from communities and progressive organizations that have tirelessly worked to expose the flaws in this system and promoted alternatives many countries are now embracing. Ever-present at the WTO ministerial conferences, annual UN climate summits and other international gatherings, members of an increasingly sophisticated social movement relentlessly critique the underlying tenets of market capitalism and promote alternatives then shared with counterparts all over the world. Their clear, consistent message is that unregulated corporate power has created deep inequality and the rampant plunder of the planet. The loss of credibility of international institutions such as the World Bank and the World Trade Organization suggests their message may be getting through.

New York Times editor Farah Stockman argued that the pandemic exposed the WTO and other international economic organizations as being completely incapable of rising to such crises. At the time the WTO was created, she noted, there was an almost "messianic" belief in the ability of unfettered capitalism to improve lives around the world. But the institution that helped create the crisis of global inequality has proven incapable of dealing with it or the other major issues of our time, including climate change.[37] In 2019, esteemed American economist and Nobel Prize winner Paul Krugman wrote in *Bloomberg* that the free-market policies he espoused for 30 years are "spectacularly useless at best, and positively harmful at worst."[38]

How all this turmoil is impacting the bottom line of corporations is debated, but there are signs that the party might not go on forever. In a 2023 report called *End of an Era*, Michael Smolyansky, principal economist of the capital markets section of the US Federal Reserve, predicted a long-run slowdown in the growth and stock returns of corporations listed on the S&P 500. Smolyansky remarked that the 30-year period prior to the pandemic was exceptional for corporate profits, due to the decline in both interest rates and corporate tax rates. However, any further lowering of interest and tax rates — which President Trump has since done in his country — would have to be debt-financed. Smolyansky predicted that stock price performance in the coming years would be under 2%, down from the average of 5.5% from 1989 to 2019.[39]

The great awakening

Parallel to this shakeup of the old world economic order has been the rise of a movement to tackle the environmental crisis. There has been an awakening, often led by Indigenous Peoples and their teachings, about the urgent need to protect nature if we are to survive. Alongside the pressing need to transition to climate-friendly energy alternatives, there is a growing understanding that humans must stop seeing nature as a vast resource for our convenience and profit, and see it instead as the very life source of our planet's existence.

In 2021, the UN published a major report saying that the world needs to quadruple its annual investment in nature if the planet is to have a future. Investing just 0.1% of global GDP every year in restoring agriculture, forests, pollution management and protected areas would close a US$1.1 trillion funding gap by 2050 and avoid ecological collapse. The UN's 2022 *The State of Finance for Nature* report said a total investment of US$8.1 trillion was required to maintain the biodiversity and natural habitat vital to human civilization.[40]

Movements to protect local land, water and forests are gaining momentum in many parts of the world. Youth climate activists are taking their governments to court for not protecting their futures from climate chaos. Many First Nations are using the United Nations Universal Declaration on the Rights of Indigenous Peoples to claim their rightful stewardship to protect the lands and waters in their territories. And all over the world, people and organizations are joining the exciting movement for the rights of nature and taking rivers, forests and lands under local community protection.

Governments and international institutions reflect these views in important venues. Recent international gatherings have emphasized the need for radical protection and restoration of watersheds, forests, wetlands and biodiversity, as well as funding for the restoration of ecosystems upon which all life depends. While many nations continue to fall behind on their commitment to the conversion from fossil fuels to renewable energy, nevertheless, excluding President Trump, the highest levels of leadership have widely accepted the need to do so.

At 2021's Climate COP26 gathering, 40 countries agreed to phase out coal-fired power altogether. At 2024's COP28, nearly 200 countries agreed to transition away from all fossil fuels. In 2023, the United Nations launched a landmark legally binding marine biodiversity agreement that would cover two-thirds of the planet's oceans outside national jurisdictional boundaries. Momentum is growing for a moratorium on deep-sea mining, with 27 countries already pledged. Over 100 countries have pledged to end deforestation by 2030, having committed billions toward forest protection and tree planting. And the member states of the UN have committed to creating a binding treaty on plastics.

Importantly, as of spring 2025, 196 nations had ratified the Convention on Biological Diversity, an international legal instrument to implement the 2030 Agenda for Sustainable Development, which, among other promises, calls for US$30 billion per year to flow from developed to developing countries by 2030 for biodiversity restoration.

At the October 2024 Biodiversity COP in Colombia, ministers from around the world declared that climate and biodiversity can no longer be treated as independent issues if either crisis is to be resolved. Calling nature and climate "two sides of the same coin," Colombia's environment minister, Susana Muhamad, said, "There is a double movement humanity must make. The first one is to decarbonize and have a just energy transition. The other side of the coin is to restore nature and allow nature to take back its power over planet Earth so that we can really stabilize the climate."[41]

These UN gatherings tend to bring together those in government and civil society who seek a regulatory framework for these crises. Their commitment to protecting biodiversity and transitioning to clean energy is genuine, as is their commitment to creating an economy that puts nature at its heart. That these wonderful declarations, pledges and commitments do not result in the action we need can be attributed to the financial realpolitik that governments face back home as well as the corporate lobbies objecting to real action or, more recently, appropriating these commitments to their benefit. The move to a nature-based economy is a perfect example of a well-intentioned narrative ripe for takeover.

The International Union for Conservation of Nature coined the term nature-based solutions two decades ago and developed global standards for governments and international institutions. Nature-based solutions, says the organization, act to protect, sustainably manage and restore nature and harmed ecosystems, benefiting nature and people at the same time. Their stated intent is to provide resilient ecosystems and ecosystem restoration, thus harnessing nature to fight climate chaos. Examples include mangrove restoration along endangered coastlines, green roofs, gardens and parks to reduce the effects of urban heat islands and sponge cities to capture rainwater and control stormwater.

Proponents point to studies such as a 2021 one led by scientists at England's Cambridge University and the Royal Society for the Protection of Birds, which analyzed 24 sites on six continents and found that returns of the "ecosystem services" created by conservation work were greater than man-made capital, created by using the land for activities such as farming and forestry. Natural sites are more valuable 42% of the time when left as they are, found the scientists, in the largest study of its kind to date.[42]

In Canada, a 2021 study reported in *Canada's National Observer* showed that ancient trees in the Fairy Creek region of Vancouver Island are worth considerably more left standing than logged. Protecting these old-growth forests could add another US$30 million in benefits to the local First Nations in the form of carbon sequestration, recreation, tourism, coho salmon habitat and plant harvesting.[43] An April 2025 study by an environmental consulting firm determined that if all the old-growth forests in regions around Okanagan and Prince George in British Columbia were fully protected, the carbon storage alone would keep the carbon equivalent of what would be generated by burning 63 million barrels of oil (over 10 billion litres) out of the atmosphere and be worth US$22 billion to the region.

"The value of the forests of British Columbia are much, much, much broader and bigger than just timber, which we have, traditionally, always and only looked at," ecologist Rachel Holt told *Canada's National Observer*.[44] On his foundation's website on nature-based solutions, renowned Canadian environmentalist David Suzuki writes, "Ultimately, we must work with nature to prevent and adapt to problems such as flooding, water scarcity, wildfires and climate disruption. When we work against nature, we work against ourselves."[45]

Nature-based solutions have become the cornerstone of the European Commission's climate plan and have been built into international climate and biodiversity gatherings, including the March 2022 fifth United Nations Environment Assembly, Climate COP27, the February 2024 Kunming-Montreal Global Biodiversity Framework (which pledged to raise US$700 billion a year to reverse the loss of nature within a decade) and the Ramsar Convention on Wetlands.[46] Between 2012 and 2021, the World Bank funded over 100 nature-based solution projects valued at US$5.5 billion.[47] The program has accelerated since then.

A report prepared by the UN Environment Programme, the World Commission on Protected Areas and others for the October 2024 Biodiversity COP in Colombia found that many countries were on target to meet the 30 x 30 goals of expanding the global network of protected and conserved areas to 30% by 2030. *The Protected Planet Report 2024* said that the global coverage of protected terrestrial and inland waters was 17.6% and of marine and coastal areas, 8.4%. While much remains to be done to achieve the goal, 51 countries have already established networks of protected areas exceeding the 30% target.[48] The European Union and the US (pre-Trump) both pledged over US$1 trillion to meet their climate promises.

The national climate plan of most governments now includes a detailed commitment to nature-based solutions. There are also expensive projects to meet government commitments to their climate goals: projects to transition away from fossil fuels, such as electric cars and home heating, public transport and carbon sequestration; to protect dwindling clean water with water recycling and watershed restoration; to plant the promised new forests and protect threatened wild spaces. The European Union and the US (pre-Trump) both pledged over US$1 trillion to meet their climate promises.

Much of the impetus to create nature-based solutions and a green economy is grounded in a genuine desire to protect nature. In a capitalist society, it is argued, we have to put a dollar figure on something for it to be considered worthy. But in adopting this narrative, the UN and its member states have unleashed great pots of money ripe for the taking. These commitments are made by governments and international institutions like the UN and the World Bank, and when they are followed up on, and the large funds available from international conservation organizations are factored in, huge amounts of public money become available to address the urgent environmental crises we face.

Financial markets grab

This has not gone unnoticed by corporations and investors and provides the newest — and perhaps the last, given the state of the planet — opportunity to plunder the commons. While carbon pricing and trading have been around for several decades, the financialization of all aspects of nature is the most recent development in this history of resource grabbing and exploitation. What became a new vehicle for corporate profit started as an attempt to address the many forms of biodiversity and environmental degradation at the highest level of leadership.

Much of the initial impetus to apply a financial model to nature was grounded in a genuine concern for the environment, part of the climate awakening. For decades, environmentalists and scientists have been touting the virtues of ecosystem services in an attempt to get governments to enact environmental protection. They reasoned that, in a market economy, a market solution to protect nature was required. If we can prove that nature has a concrete monetary value, it can compete in the open market in its natural state against the other uses to which the forests, watersheds and land might be put.

But this approach started a dangerous trend. US-based Food & Water Watch says that the process of bringing nature under the control of the "logic of the market" happens in stages. First, it is made a commodity — the commercialization of something not generally seen as a product. Commodification turns an inherent value into a market value, enabling it to be bought and sold. Privatization transfers control and management of commoditized resources from public ownership to private ownership. The commodities can then be priced, and a market can be created for them. At this point, financialization acts upon the commodity as an asset and applies various financial instruments to it, such as water futures contracts or carbon credit options.[49]

In 1997, a group of American ecologists and economists teamed up for the first attempt to put a price on nature. In the journal

Nature, the experts stated that, every year, the world's diverse ecosystems combined produced on average US$33 trillion worth of ecosystem services. For context, they noted, worldwide gross national product that year was around US$18 trillion.[50]

The report was met with caution by some conservationists. Dr. David Ehrenfeld of Rutgers University told the *New York Times* that common sense and "what little we have left of the wisdom of our ancestors" tells us that if we ruin the Earth, we will suffer grievously. "I am afraid that I don't see much hope for a civilization so stupid that it demands a quantitative estimate of the value of its own umbilical cord."

But a number of senior scientists and economists said that the report laid the foundation for a whole new way of looking at and protecting nature. Columbia University economists Drs. Graciela Chichilnisky and Geoffrey Heal used the report to call for selling investment shares in ecosystems. Using the Catskill watershed as an example, they wrote that the capital thus raised would pay for preserving the watershed. "Returns to investors would come either from a share of the costs saved by not having to build a treatment plant or, if the investment were private, by actually selling ecosystem services. In the case of a watershed, clean water would be sold."[51]

In 2001, the UN launched the Millennium Ecosystem Assessment, a four-year study involving more than 1,300 leading scientists from 95 nations that provided detailed scientific documentation of the planet's ecosystems and the services they supply. They grouped ecosystem services into four categories in an attempt to quantify nature's contribution to our well-being and health. Ecosystems provide us with food, clean air and drinking water; they regulate climate and help control disease; they support nutrient cycles and crop pollination; and they provide us with cultural, spiritual and recreational services.

The World Resources Institute said in 2005 that the report offered the first truly comprehensive picture of the health of the planet. "The Millennium Ecosystem Assessment provides an indispensable baseline of information for researchers, scholars, and students, as well as informs public decision-making for decades to come."[52]

Although this report provided much badly needed information on the planet's ecosystems and documented the harm to nature caused by human activity, the framework for this in-depth analysis was entirely human-centred and basically ignored the ecological damage to nature and biodiversity itself. In fact, by framing nature as providing "services" to humans, the UN set the stage for others to view nature as a market commodity and put a dollar figure on it.

At the 2012 Rio+20 Summit held in Brazil, the Nature Conservancy and Corporate EcoForum as well as 24 Fortune 500 companies, including Coca-Cola, Weyerhaeuser, General Motors and Dow, released a report based on the findings of the UN undertaking. Called *The New Business Imperative: Valuing Natural Capital*, the report estimated that the world's ecosystems contributed about US$72 trillion in goods and services every year to the global economy.[53] As with many other "market goods" (food, housing, energy), the price of a marketized nature continues to grow. In its 2018 *Living Planet Report*, the World Wildlife Fund estimated that nature was now providing services to humans worth US$125 trillion a year.[54] In a 2023 report, The World Wildlife Fund wrote that water's "ecosystem services" alone contribute US$58 trillion annually to humans.[55]

The corporations involved in the Brazil gathering clearly laid out their blueprint as the Earth's population was starting to truly grapple with the enormity of the climate crisis. Their message was not subtle. The private sector has the power to "leapfrog over government inertia," said the report, giving private companies a "new business imperative in ecosystem management." Corporations, they argued, stand to benefit from prioritizing ecosystem services in four ways: reducing risks, cutting costs, enhancing a brand and growing revenues.

The 24 corporations that signed this document clearly stated their intentions: "Exploit opportunities to educate consumers about the high-performance products to increase demand and create new market segments . . . Leverage emerging 'natural capital' markets such as water-quality trading, wetland banking and threatened species banking, and natural carbon sequestration." Communities are called "customers" and corporations are urged to position themselves for long-term success by "putting a monetary value on what nature does for their business." Nature is hardly mentioned in the report and only as a backdrop to the key economic message.[56]

Green capitalism grows deep roots

Over the next decade, with global attention shifting to biodiversity protection and other environmental priorities related to the climate crisis, private corporations and asset managers took a leading policy role through promoting the financialization of nature. In 2016, more than 50 transnational corporations joined with Conservation International to create the Natural Capital Coalition and a Natural Capital Protocol to provide businesses with a standardized way to incorporate "natural capital" into their plans. They found support at the UN, the World Bank, the European Commission, governments of the Global North, the OECD, major conservation organizations and other centres of power.

Unfortunately, the United Nations, while having pioneered so much international commitment to tackling the climate crisis through promoting clean energy and protecting biodiversity, has embraced carbon trading. It has set up the United Nations Carbon Offset Platform, "an e-commerce platform where a company, an organization or a regular citizen can purchase carbon credit units to compensate greenhouse gas emissions."[57] The UN Environment Programme publishes an annual *State of Finance for Nature* report and holds Investing with Nature webinars to help financial institutions invest in nature-based solutions.

Three decades ago, the World Bank and the United Nations created the Global Environment Facility (GEF), a family of funds that has provided more than US$25 billion in financing for conservation and biodiversity protection. More recently, GEF has wholeheartedly adopted the natural capital framework. The World Bank says that with external financing from GEF, it offers "blended and flexible financing options, often serving as the initial funding that facilitates larger projects and attracts additional private investment. In this way, it has played a key role in de-risking investments, piloting innovative solutions, and mobilizing private capital for climate and nature."[58]

Overall, the World Bank has taken a very strong stand in promoting the financialization of nature. In reports such as *Unlocking Nature-Smart Development* (2021) and *The Economic Case for Nature* (2021), the bank clearly states its goals for future financing. The Washington-based World Bank Group is the largest development bank in the world, providing around US$100 billion in loans and assistance to developing countries every year. A stated goal of the group is to strengthen the private sector, including financial institutions, in developing countries.

"The World Bank Group's work on biodiversity goes beyond conserving and protecting nature. Biodiversity is integrated into the broader portfolio of the World Bank in sectors like agriculture, forestry, watershed management, fisheries, and coastline zone management," it says on its website. It promotes mobilizing finance through the design and application of financial instruments for nature such as "labeled bonds, transition bonds, sustainability-linked bonds, and insurance products." Over the last decade, the group has created a US$750 billion green bond market "connecting environmental projects with capital markets and mainstream investors."[59]

The bank has come under serious criticism for shoddy data collection and reporting on its green finance projects. In a 2024 report, Oxfam International found that up to US$41 billion in World Bank climate finance — nearly 40% of all climate funds disbursed by the bank over the preceding seven years — is unaccounted for due to poor record-keeping practices. Oxfam's investigation revealed that obtaining even basic information on how the World Bank is using climate finance was painstaking and difficult. "Not tracking how or where the money actually gets spent? That's not just some bureaucratic oversight — that's a fundamental breach of trust."[60]

Rich countries jump on board

Nations of the Global North have embraced the language of a nature-based economy while assuring it poses no threat to business as usual.

In Great Britain, a multiyear report on the economics of biodiversity commissioned by the UK Treasury and led by economist Sir Partha Dasgupta was published in early 2021 to much fanfare. It was widely reported on and understood to have set the 25-year plan for environmental policy in Great Britain. The launch was attended by the Prince of Wales and the prime minister and endorsed by the famous broadcaster and naturalist Sir David Attenborough. The report clearly sets out a nature-based solutions approach to environmental policy, presenting biodiversity loss as a problem of asset management, and asserting that economic growth is compatible with sustainable development. While no doubt the authors of this landmark report genuinely care about the climate crisis, in the end, they could not see a way forward without involving the market.

Under President Biden, the aim of the United States' Green Growth Strategy was to facilitate economic growth by emphasizing efficient use of natural resources. In a 2021 publication, the US Department of State recommended that consumers can help develop a green economy in their buying habits and praised the TV show *Shark Tank* for completing successful deals related to insect protein snacks and reusable dry-cleaning bags.[61] Biden's 2021 bipartisan infrastructure deal announced US$1 trillion for climate mitigation through providing clean water, building public transportation and electric vehicles, and tackling legacy pollution. Not only would this fulfill the country's climate commitments, the government boasted, it would be a boon to investors and the economy by opening up new business opportunities such as public-private partnerships.

Under Donald Trump, of course, this project and many other climate-related commitments were cancelled, and a number of companies and banks withdrew theirs as well. Some observers speculate that corporations and banks are now off the hook on climate and no longer have to claim to have green ambitions. But in a January 2025 article, the Bureau of Investigative Journalism noted that green pledges by banks and asset managers were in response to pressure from shareholders concerned about the climate crisis and they abandon those investors at their peril. As well, renewables are now the cheapest form of energy generation and investments in wind and solar are expected to surpass fossil fuels in the next several years.[62]

Canadian reporter Max Fawcett wrote in 2025 that even Donald Trump cannot slow down the energy transition. He pointed to reports showing that every major renewable energy technology decreased in cost in 2024, with battery storage down one-third and solar farms down 21%. Countries such as Saudi Arabia, the United Arab Emirates and China are investing heavily in alternative energy technology, the latter on track to wean itself off fossil fuel imports in the foreseeable future. Global clean energy investment was worth more than US$2 trillion in 2024, but it will need to more than double to meet the Paris climate goals.[63]

Federal US spending is not the only source of green finance in that country. Many states sponsor "green banks" that attract and leverage billions in private capital to promote clean energy projects. In 2024, California's green bank invested US$25 million to finance sustainability goals "while simultaneously creating good jobs and promoting economic development in Californian communities." In 2024, the State of California became the sole investor in a World Bank five-year US$150 million green bond issue to support the funding of "climate resilient growth."[64] The good news is that even Trump cannot stop the move toward a clean energy future in

California and some other states. The bad news is their investment is linked to and dependent upon a corporate-dominated model.

Nations of the Global North have also embraced the market opportunities of this growing commitment to biodiversity protection as being an exciting new opportunity for their industries and exports. But while they are signing pledges at UN conventions to protect biodiversity and water and curb their fossil fuel emissions, they are promoting their domestic companies and resource development.

The Trade and Investment Commission of the Australian government says the "thriving" green economy is "transforming international trade" and boasts that Australia is "well positioned" to be an "important partner" in the transformation of the global economy. It points out that the country has a competitive advantage with its abundant natural resources and cites the demand for the export of its critical minerals.[65] In 2024, the Australian government issued its inaugural US$4.5 billion green bond, calling it a "golden opportunity" to make Australia a "go-to destination for international green capital."

The Canadian government is proud of its plans for a "green future," wrapping them in the language of clean fuels, clean energy, clean electricity and green jobs. But even as it touts its green credentials in the forestry and energy sectors, Canada is shipping wood for biomass export from old-growth forests and has steadily increased oil and gas production and exports from its infamous tar sands operations. Much like other countries, stated international commitments and pledges Canada has made are often not met or delivered in a way that does not threaten domestic corporate interests.

The private sector cashes in

Meanwhile, private sector participation in the green economy has exploded. The UN Environment Programme reports that private finance for nature surged elevenfold in four years, from US$9.4

billion in 2020 to US$102 billion in 2024. While the public sector is still funding the majority of nature-reaped finance, the UN says if the pace of market growth continues, this gap could be closed by 2030, with a possible private investment of US$1.45 trillion. Private expansion is taking place through "alternative investment, debt-for-nature swaps, private equity funds, nature-based Exchange Traded Funds, as well as new instruments like biodiversity credits."[66]

A consortium of European journalists and leading media outlets reported in 2022 that European investment funds labelling themselves as sustainable totalled over US$4.5 trillion, amounting to the stock market value of Alphabet, ASML, Coca-Cola, Nestlé, Pfizer, Samsung, Shell, Toyota, Walmart and Walt Disney combined. The journalists studied 838 funds that label themselves "dark green," the highest possible sustainability label, and found that half of these also invest in fossil fuels and aviation.[67] Carbon trading has hit record highs in the last several years in spite of having developed an international reputation as a false solution, since corporations take advantage of the offsets rather than lightening the actual climate footprint of their business.

According to newsletter *Net Zero Tracker*, by 2023, two-thirds of the world's largest 2,000 companies had adopted a net-zero target, and most of them are actively trading in carbon. UK-based news outlet *Carbon Brief* reports that two-thirds of the world's biggest companies with net-zero emissions targets are using carbon offsets to help meet their climate goals.[68] European carbon consultant DGB Group anticipates that the carbon offset market will soar to US$1.6 trillion by 2028.[69] And according to a 2025 forecast by *Yahoo Finance*, the carbon credit and offset markets are expected to expand at a compound annual growth rate of almost 12%, projected to reach US$2.73 trillion by 2032.[70]

In January 2024, 320 corporations and financial institutions representing over US$18 trillion in market capitalization signed up to disclose their nature-based practices and commitments with the Taskforce on Nature-Related Financial Disclosures. Business portfolios are dependent on nature and climate, said one corporate investor, and must be treated as both strategic risks and investment opportunities.

Other corporations shared the fact that regulation to protect nature is growing in a number of countries and touted the fact that the member companies and banks will be reporting to the International Sustainability Standards Board — an independent, private-sector body that develops and approves sustainability disclosure rules — establishing a voluntary process for their reports.[71]

Critics of corporate disclosure say that these corporations have only agreed to strong calls for more transparency in meeting their climate promises on the condition that the system is self-governing. Rainforest Action Network, a US-based global network fighting to protect forests, asserts that the taskforce is entirely an invention of the private sector and reflects only the interests of the corporations it claims to regulate — the very corporations driving the biodiversity crisis. It is made up of 40 senior staff from transnational corporations such as Bank of America, one of the world's biggest fossil fuel financiers, and BlackRock, the world's biggest investor in deforestation. The process promotes greenwashing, says the network, and sidelines the very communities its corporations hurt.[72]

The North-South divide deepens

This is not to say that there is not genuine commitment behind some projects based on a nature-based narrative. Germany has pledged to reduce its greenhouse gas emissions by 55% by 2030. To fulfill its biodiversity commitments, Canada is expanding its national urban parks initiative and creating massive new parks in wilderness areas, the latest being the Thaidene Nënë National Park Reserve, over 14,000 square kilometres of boreal forest in the Northwest Territories, and the Pituamkek National Park Reserve, a 30 square kilometre chain of barrier islands off the coast of Prince Edward

Island. The African Great Green Wall is underway to restore 100 million hectares of degraded land and reverse desertification of the Sahel region. The key is democratic, not corporate, control.

It is also very welcome when industry and corporations take action to clean up their own environmental footprint. Ceres, a US-based non-profit advocacy organization "working to accelerate the transition to a cleaner, more just, and sustainable economy," helped to organize more than 130 major companies and financial institutions on the eve of the October 2024 UN Biodiversity COP16 to call on governments and global leaders to accelerate the adoption and enforcement of robust nature policies. They said governments need to strengthen — not weaken — "the policies, incentives, legislation and regulation essential for encouraging ambitious and effective corporate action on nature."[73] Their Business for Nature Coalition supports banning land conversion in key protected areas, transitioning to regenerative farming models and adopting a global moratorium on deep-sea mining.

However, as much as some of these companies may be genuine in seeking good environmental regulations, the inclusion of others raises questions about greenwashing. One signatory company is British giant Unilever, criticized in a 2023 Greenpeace report for selling 1,700 highly polluting plastic sachets every second, feeding the global plastic crisis. Other signatories include bottled-water companies Danone and Nestlé, among the top five plastic waste producers in the world.[74]

Whatever the individual intentions of corporate leaders themselves, the result of the global private sector takeover of climate and biodiversity protection is a danger to the planet, to justice and to the health of the majority of the world's population. The problem with the financialization of nature, regardless of the intent of its perpetrators, is the framework it represents — that nature can best be protected by taking it away from governments and communities and bringing it into the market economy, where it will compete to survive.

Green growth also allows the current system of deep inequality between and within nations to remain intact. Most powerful transnational corporations, asset fund companies and banks leading green growth are situated in the wealthy countries of the Global North (or are among the wealthy elite of some in the Global South), and it is their millionaires who are quickly becoming billionaires, giving them incredible political influence. And most people in the Global South and many in the wealthy countries of the North have not reaped the financial benefits of these "investments in nature," increasingly enjoyed by only the elite.

American political economist John Bellamy Foster argues plans for the expropriation and accumulation of natural capital by global finance have been primarily directed at the biodiversity wealth of the Global South and cites UN documents that show just that. United Nations Environment Programme spatial mapping shows a high concentration of "terrestrial ecosystem assets" in the equatorial regions, such as the Brazilian Amazon and the Congo Basin, and the "high-value marine ecosystems" of Southeast Asia. Bellamy points out that much of the land in sub-Saharan Africa is untitled, making it vulnerable to grabbing. He fears a "grand expropriation" of the commons of the Global South and dispossession of humanity on a scale exceeding all previous human history.[75]

As reported by the political economy journal *Phenomenal World* in 2021, the emphasis on capital flows to come to the rescue of the climate has also left the structural features of the global financial system largely unaddressed, entrenching deep global inequities. Sustainable finance is booming in the Global North where environment-themed funds are on track to break records, noted the journal. But in the Global South, the story is different. The availability, cost and terms of finance to develop sustainably remain restrictive, leaving poorer countries little space to manoeuvre with international capital markets. In fact, the ascendency of "non-state actors" — especially big corporations and industry groups — in the official UN climate process has given great authority to an "alliance of asset managers and other investment institutes that are among the biggest managers of developing country debt."[76] The financialization of nature is perpetuating and deepening the North-South divide.

Ghanaian development researcher Thelma Arko says the growth of what she calls "green colonialism" is fuelling a new scramble for African land and perpetuating colonial-era exploitation. "Carbon cowboys" expropriate Indigenous Peoples and small farmers from their territories, sometimes violently. "This dispossession of ancestral lands and livelihoods not only prioritizes carbon sinks and conservation areas over the subsistence farming, pastoralism, and cultural practices of local communities, but also disrupts their way of life. . . . The imposition of conservation or renewable energy projects without the free, prior, and informed consent of local communities, as outlined in the United Nations Declaration on the Rights of Indigenous Peoples, is leading to the destruction of biodiverse ecosystems, disruption of water cycles, and loss of habitat for endangered species."[77]

Friends of the Earth International (FOE), the world's largest grassroots environmental network, has published a comprehensive series of critiques on the financialization of nature. "'Green Growth' redefines 'green,' not 'growth,'" notes FOE. Nature, especially in the Global South, is divided into different ecosystem services that can be quantified, measured and above all, broken up into individual units so profit can be made from selling rights to these units of nature. Markets in ecosystem services need clear and measurable units. But nature doesn't come with neat beginnings and endings — nature is a dynamic interaction, says the network. In their report *Financialization of Nature*, they write, "The environmental justice perspective understands that economic valuation and financialization of nature are simply the latest examples of capital markets using nature for profit maximization, as they have been doing for centuries. . . . It is such a perverse world where corporations are people and forests are bundles of carbon, water and biodiversity offsets. . . . Financialization represents further reduction of community control over their territories and an extension of the local license for corporations to destroy the web of life we depend on and which is showing increasing signs of multiple crises."[78]

It turns out that green capitalism and the commodification and financialization of nature is the latest — and perhaps last — enclosure of the commons.

CHAPTER TWO

FOLLOW THE MONEY

If the soil is washed off the land, we cannot grow crops on a bed of derivatives.

— GEORGE MONBIOT, *THE GUARDIAN*

Rolando Zumba, a gentle 59-year-old, wept through the Associated Press (AP) interview as he described the displacement of his people and their traditional lands to make way for a giant carbon offset project. His own ability to hunt was ended when rangers in Peru's Cordillera Azul National Park — a spectacular 13,500-kilometre sweep of Peruvian Amazon rainforest, mountains and waterfalls in the foothills of the Andes "where clouds cling to the treetops and morning mists settle over powerful rivers" — confiscated his hunting rifles, ending self-sufficiency for his Kichwa tribe on its ancestral land, ensuring poverty and hunger for his people.

In a December 2022 exposé, the news agency reported that this project was flawed from the beginning. The carbon credit program was set up in 2008 to "offset" the carbon footprint of Shell and TotalEnergies, who purchased blocks of the park, allowing them to claim more than 28 million "credits," meaning they were theoretically offsetting that much CO2. The project brought in so much money, it covered around 90% of the operating expenses of the park and was supposed to be used for forest protection and reforestation. But the AP exposé found that not only did the companies exaggerate the credits earned, tree cover loss in the park also dramatically increased due to increased deforestation.

The story of the displacement of Indigenous and ancestral dwellers from their traditional lands to make way for "nature-based" biodiversity protection areas is distressingly common, often the cruel legacy of clearing lands for ecotourism and resource development. It is a further violation of human rights when major energy companies seek to offset their own carbon footprint in these "protected areas" by offering up large sums of money to cash-strapped governments of the Global South to maintain them. The Kichwa people had lived in these forests from a time before memory. Now, reported AP, the tribe is so poor, they cannot send their children to school. "We live like dogs," said one Kichwa mother.[79]

How the financialization of nature works

The market approach to nature is now deeply entrenched in many governments and international institutions, and trading in nature's "assets" is a huge business. Most people around the world know little or nothing about this fast-moving development. But its reach is astounding.

The financialization of nature is modelled on the same financialization of the economy that promotes investments, not in tangible things such as goods and services, but in risky financial transactions that attempt to profit from the fluctuations in the market value of assets. Trading money and speculating become more profitable than producing goods or providing services. It is all about risk, and the winners are those who most successfully predict the future. Through these devices, financial markets, financial institutions and financial elites gain great influence over economic policy and government authority. This, in turn, limits the role of governments and local communities in protecting their own interests, including human rights or the environment.

Nature-based finance operates on the same premise, but zeros in on climate change and other environmental crises. While it might not be possible to predict exactly where and when disasters will strike, those in the risk business can safely bet that the climate crisis will continue to provide them. Every firestorm, hurricane and flood presents an opportunity for experts in risk forecasting to advise on the clean up and recovery technologies that will be needed. And governments give large tax breaks and other incentives to the corporations, banks and asset management companies that use "green" funds to underwrite such businesses.

Crucially, green finance is situated within the mainstream of traditional economics whose orthodoxy is not to be questioned. Dealing with the climate crisis in a way that does not challenge the economic status quo requires an adherence to the dictates of the current market framework. Political leaders turn to experts in the workings of the market economy to guide them to net zero with as little disruption to the economy as possible. Painted with the brush of objective science and rationality and enveloped in the mythology that there is no alternative, mainstream economics continues — with great effect — to cleanse our ecological crisis of its political origins and resolutions.[80]

Green finance is part of the doctrine of neoliberal capitalism, which holds that the market can resolve almost all social, economic and political problems, including the climate crisis, better than governments. George Monbiot explains in *The Guardian*, "I'm talking about the development of what could be called the Natural Capital Agenda: the pricing, valuation, monetization and financialization of nature in the name of saving it."[81]

Well-meaning investors fuel green finance

In many ways, well-meaning public opinion is stoking this development. The private sector has been the beneficiary of an increasingly informed global public demanding that something be done to confront the climate and other environmental crises. There is a deep desire on the part of many individuals and institutional investors, such as pension funds, university endowments and non-profits to invest ethically, with a strong emphasis on protecting nature and promoting nature-friendly technology. Polls show a growing citizen awareness of the climate, water and biodiversity crises, as well as demands for governments to take action.

Four in five Canadians "of all political stripes" want stronger government action to protect forests and wildlife.[82] Two-thirds of Americans think their government should do more on climate.[83] Over 80% of Britons supported solar, water, tidal and wind power in a 2024 survey.[84] However, a summer 2025 poll found other urgent issues are overtaking the climate for many in Great Britain.[85] More than eight in ten EU citizens want their governments to invest "massively" in renewable energies.[86] Nine in ten Chileans and eight in ten Brazilians demand stricter climate policies from their governments.[87] One-third of India's people worry about climate every day, and almost 80% want more government action.[88]

A major June 2024 global survey carried out by the UN and the University of Oxford found that 80% of people globally want stronger climate actions by their governments. The biggest ever standalone public opinion survey on climate change found that even more — 86% — want to see their governments set aside geopolitical differences and work together on climate change.[89] And a global survey of 130,000 people across 125 nations found a "silent majority" in every country wants their government to do more to fight climate change. Published in April 2025, researchers from a number of universities, including Yale and the University of Bonn, said that, although an overwhelming majority among the world's people want stronger action to fight the climate crisis, they are trapped in a self-fulfilling "spiral of silence" because they mistakenly believe they are in a minority.[90]

So it is no surprise that the same individuals and institutions asking their governments to step up will seek out environmentally ethical ways to invest their money. Many people and organizations who may be critical of Big Oil, for instance, are happy to embrace the notion of a green economy, with its "ecosystem services" and "nature-based solutions." They believe it is time the economy was geared to protecting nature (as do we all) but perhaps have not yet been exposed to the ways in which many corporations are using the climate crisis and the financialization of nature as an opportunity to greenwash their operations.

And that poses a serious problem. As Adrienne Buller writes, "'Green' or 'ethical' investing can far better be understood as a means through which the wealthy and those with financial assets can bet on the likelihood of a greener economy, rather than contribute to bringing that economy into being."[91]

Asset management companies becoming powerful players

Sustainability-linked stocks and bonds are appealing to ethical investors, and companies are competing for those funds by wrapping themselves in promises to be part of the climate solution. Green financing serves the purpose of allowing people to feel good about investing in the market while not challenging the deep flaws in the increasingly deregulated market economy and the part it plays in nature's destruction. Investments in funds marketed as ESG (Environmental, Social and Governance), three areas identified for evaluating "good" corporate behaviour, are skyrocketing as companies and asset managers compete to look environmentally responsible.

In 2021, *Bloomberg* reported that over 40% of assets under management (the amount of capital managed by investment firms on behalf of their clients) already claim adherence to ESG principles and that 85% plan to increase their commitment to an ESG framework in the future. *Bloomberg* stated that ESG assets reached US$30 trillion in 2022 and predicted they would surpass US$40 trillion by 2030.[92]

This growing environmental consciousness, particularly among young investors, is hugely profitable and has helped to explain the meteoric rise in the asset management industry.

In asset management, financial companies manage money on behalf of institutions, sovereign wealth funds, corporations, pension funds and others. They index their portfolio to the stock and bond markets, and investors trust them to make financial decisions on their behalf. Asset managers have become major shareholders in thousands of corporations and strongly influence their policies and actions. *Nonprofit Quarterly Magazine* reports that the industry, especially the Big Three (BlackRock, which *Bloomberg* calls the "fourth branch of government," Vanguard and State Street), has seen "astronomical growth" in the last 15 years. The asset management industry is replacing traditional banking with assets under management worth well over US$100 trillion — around double the 10 largest national GDPs in the world combined.

Because of their large stake in the global economy, reports the journal, asset managers play a powerful role in driving the crisis of modern capitalism, including the climate crisis. The Big Three manage a huge stake in the fossil energy industry — over US$300 billion in listed fossil fuel investments through their funds alone — while promoting green capitalism as the answer to the climate crisis.[93] In *The Value of a Whale*, Adrienne Buller notes that through their sheer scale, a small cohort of asset managers has come to have a steering role in advising governments and international institutions on how to confront the ecological crisis. They place bets on the future while contributing to its shape. And their wealth has allowed them to emerge unscathed, even in times of crisis for people and the planet.[94]

Yet even these powerful players are facing the volatility of an unpredictable market and political turmoil. As noted earlier, the global trading system had already flatlined even before Donald Trump launched his trade wars, and there are many uncertainties. The international consulting firm Protiviti reports that inflation, market volatility, regulatory compliance and changes in the interest rate environment are major recent concerns for asset manager CEOs.[95] "Staring down the barrel of an unprecedented threat to returns and asset values," writes Buller, "the new titans of finance are undertaking a profound effort to ensure global efforts to tackle ecological crisis align with their interest, and those whose wealth they manage."[96]

Corporate-connected conservation NGOs are powerful players as well

These asset managers and the World Bank work closely with several influential and well-funded international environmental and conservation organizations that are leading voices in promoting the financialization of nature. They have been criticized by many for helping to "greenwash" the profits of the finance industry. Transnational Institute, a Europe-based research and advocacy organization, is deeply critical of the promotion of "conservation finance" by leading NGOs.

"The basic premise of conservation finance is that saving nature and averting the climate crisis requires enormous funds, but money derived from public and philanthropic grants is woefully insufficient. Proponents argue that the only way to bridge this funding gap is to tap into the trillions of dollars of private capital circulating through global financial markets. To do this, saving nature must be turned into a profit-making endeavour, appealing to what are known as 'impact investors,'" says the organization.

"The rise of conservation finance has transformed not only the way in which conservation is addressed, but by whom. People with backgrounds in finance, banking and business consulting are taking over the management of most of the big conservation organizations. Their governing boards are stacked with investment bankers, hedge fund managers and venture capitalists. Consequently, risky and opaque financial instruments, originating in financial markets, are being repurposed for environmental projects. . . . This process represents another dimension of financialization; the process whereby financial markets, financial institutions, and financial elites are gaining greater influence over almost all aspects of society."[97]

This is a concern as some conservation NGOs are taking over the role of governments in owning and managing what were once government-funded protected areas, such as national parks. As government funding dries up, these well-heeled organizations are able to step in to fill a funding void. Land managed or owned by NGOs has grown at a "staggering" rate in Australia, according to a study by the University of Queensland, and the country now has the second-largest percentage of land managed privately for conservation in the world, after the United States.[98] Many individual and institutional investors are happy to support projects and NGOs that buy up land to protect it and its biodiversity. But they might think twice if they realized their investments were going to support the interests of some of the world's leading polluters.

The Nature Conservancy, the world's largest environmental organization with total assets of US$9 billion, has helped to protect huge swaths of land and billions of hectares of marine habitat. But the organization has long been criticized over its close relationship to some of the biggest corporations and investment firms on the planet. Board members have included CEOs from Alibaba, Duke Energy, China Capital Group, the Carlyle Group, BlackRock, J.P. Morgan and Hewlett-Packard.

Politico reporter Zack Colman wrote that Mark Tercek, who came from Goldman Sachs in 2008 to run the TNC until his departure in 2019, brought with him "Wall Street–style practices." In 2013, Tercek published *Nature's Fortune: How Business and Society Thrive by Investing in Nature*. He set up an investment project for the conservancy called NatureVest and partnered with corporations that had reputations as environmental offenders, including Coca-Cola, oil giant BP, mining heavyweight BHP Billiton and Dow Chemical.

Colman reports that CEO Tercek leveraged his investment banking background and a partnership with JPMorganChase to drive US$1 billion of private capital for conservation investments, including carbon credits. He also hired former investment bankers and added alumni from big agriculture companies, including Monsanto.[99]

NatureVest's website spells out its mission clearly. "Private investment and capital markets have a crucial role to play in addressing the climate and biodiversity crises. We are dedicated to developing a pipeline of impactful investments and broader market interventions that can deliver conservation at scale. . . . Our impact investments and market innovation strategies focus on forestry and carbon, water and agriculture, fisheries and aquaculture, infrastructure and renewable energy, sovereign debt solutions, and conservation tourism."[100]

In 2023, The Nature Conservancy launched a Nature Bonds Program to work with governments of the Global South to refinance debt, generate new funding for investments in conservation and fight climate change. This is the newest iteration of debt-for-nature swaps that involve some of the largest corporate investors in the world, who then get not only significant control over policies of the debtor country but must find a way to profit from their investments. TNC is promoting nature bonds in spite of the fact, as reported by the *Financial Times*, it has dropped its "blue bonds" program for marine conservation as bonds issued to finance a deal with Gabon were used to help the country refinance other debt rather than being used for conservation.[101]

World Wildlife Fund (WWF) is another highly influential environmental organization that promotes nature-based solutions and ecosystem services. It calls itself the world's largest conservation organization, working in over 100 countries. Like The Nature Conservancy, WWF has long been criticized for its close alliance with big corporations. A 2014 book by German author Wilfried Huismann exposed that the WWF has accepted large donations from Coca-Cola, Shell, Monsanto, HSBC, Cargill and BP, among others, companies that benefited from the organization's green image to counter criticism of their environmental footprints. The WWF has an elite club of 1,001 wealthy members who advise the organization on projects and policy.[102]

WWF recently announced it is setting up the Nature-Based Solutions Origination Platform to create a new model of scaling up private investments in "high-quality" projects. One could be forgiven for finding it difficult to understand the real purpose of the project from its description: "The NbS-OP will generate a supply of high-quality landscape investments . . . and link it with diverse funding sources. The platform aims to showcase high-quality interventions that address threats and drivers efficiently; incorporate transparent and equitable governance and benefit-sharing mechanisms; and generate durable impacts for climate, biodiversity, and sustainable development."[103]

The World Wildlife Fund has called for nature-based solutions to be included in the Convention on Biological Diversity's Global Biodiversity Framework. In 2022, the WWF teamed up with the World Economic Forum to promote a "nature-positive economy" so that upcoming COPs on climate and biodiversity would include nature-based solutions in all their commitments. The World Economic Forum is a powerful international organization that promotes public-private partnerships and meets in Davos, Switzerland, every year to bring senior politicians and business leaders together. While the language of their campaign may sound as if these organizations are all about stopping biodiversity destruction and addressing climate change, the WEF and the WWF call for "nature markets" to be embedded in the definition of a nature-based economy.

Measuring creation

Key to establishing nature-based markets is collecting data on nature — the forests, fields, wetlands and wildlife that make up our natural world. As with the first nature data collection project, the 2001 UN Millennium Ecosystem Assessment, much of the impetus for this research is driven by good intentions. Some governments and communities want to place a value on nature in order to protect it, which means first collecting as much information as possible on their natural bounty. But this accounting is also a necessary precondition for those corporations and investors who would turn nature into a marketable commodity.

The official term used is "natural capital accounting," cold language reminiscent of financial accounting. The European Commission (EC) defines natural capital accounting as "a tool to measure the changes in the stock and condition of natural capital (ecosystems) at a variety of scales and to integrate the flow and value of ecosystem services into accounting and reporting systems in a standard way. . . . Natural capital accounting has the potential to provide a concrete basis for business performance reporting by explicitly mapping out impacts and/or tendencies on natural resources and placing monetary value on them."

In 2022, the EC developed standardized natural capital accounting principles for businesses to align with the European Green Deal. Called Transparent, the project is supported by the World Business Council for Sustainable Development and was created because "the business community is calling for holistic datasets and standardized methodologies that allow them to include the value of nature in their internal decision making."[104] Not a word about nature protection on this site.

In 2021, the United Nations established the System of Environmental-Economic Ecosystem Accounting, standardizing ecosystem accounting in response to the call in Agenda 21 that the values of nature be recognized within all the information systems of member states and all relevant UN agencies. The UN system is an internationally agreed-upon statistical framework that describes ecosystems and the services they provide to people and the economy. Under this accounting system, ecosystems are considered to be assets.

Five countries were chosen to pilot this project, co-sponsored by the European Union, UN Environment Programme and the Convention on Biological Diversity. Organized by South Africa, the Africa Natural Capital Accounting Community of Practice was created in 2019, and the continent now holds an annual Global Policy Forum on Natural Capital, whose mandate is "to connect users — primarily the investment community and policy makers — with both data and analytics, and data providers, to share knowledge and experiences on incorporating natural capital accounting and approaches in decision making."[105]

On Earth Day, 2022, former US President Biden signed into law a national strategy to "reflect natural assets on America's balance sheets," leading to a nationwide natural capital accounts project. "These natural capital accounts will measure the economic value that natural resources provide to society and illustrate how a robust economy depends on a healthy natural environment," said the press release. The government will develop statistics for environmental-economic decisions, to "reflect natural assets on the national balance sheet . . . highlight a framework for investment opportunities . . . and take stock of our wealth of natural assets."[106] While the project came under almost immediate attack from American free enterprise adherents after Donald Trump took power, it was well underway by then and could still be championed by a future administration.

The pressure is on the Canadian government to measure and price its vast biodiverse natural wealth. In 2024, Canada recently created a Census of Environment, based on the UN's accounting system, to provide "measures of ecosystem services such as food, clean air, clean water, carbon storage, wildlife habitat, and recreational opportunities."[107] Already, dozens of towns and cities have adopted nature-based accounting in their financial plans and in 2025, the University of Waterloo, working with KPMG, one of the largest professional accounting firms in the world, created a guide to standardize the practice across the country.

Australia has several private natural capital companies taking advantage of the country's financial commitment to measuring its ecosystem services. *Accounting for Nature* promises to "make nature count," and says on its website, "You can't manage what you haven't measured."[108] In 2024, the US, Canada and Australia signed the "Partnership on Cooperation on National Capital Accounting and Environmental Economic Accounting," recognizing that "natural resources are capital assets critical for economic growth and prosperity" and pledging to link "natural capital accounts" to "national economic accounts."[109]

China is pioneering a new ecosystem measurement and accounting system called the Gross Ecosystem Product. The Chinese Academy of Sciences is partnering with Stanford University's Natural Capital Project, whose mandate is to "improve the well-being of all by motivating greater investment in natural capital." The world's ecosystems can be seen as capital assets, according to Stanford's website, which touts "the measurement of ecosystem assets as stocks and ecosystem services as flows" and adds, "Gross ecosystem product is a measure of the aggregate monetary value of final ecosystem-related goods and services in a specific area and for a given accounting period."[110]

From the perspective of those who care about and for the natural world, this increasingly cold and bureaucratic language to describe the process of collecting data on nature is deeply alienating. The consistent message of the financialization of nature is that nature exists primarily to serve humans and human well-being. Terms like "ecosystem services" and "nature-based solutions" place the natural world at the service of humans and our inalienable right to dominate the Earth and other species. The notion that plants and forests and rivers and animals could have rights outside their material use to us is non-existent in most of the sophisticated publications of those profiting from greening the economy.

Green bonds fail the climate

"Green bonds," also called "climate bonds," are repayable loans to governments, banks and corporations that promise to be used for positive climate change or environmental projects. Green bonds follow the Green Bond Principles laid out by the International Capital Market Association, a global trade association for companies in the cross-border debt capital markets. Green bonds have been around since 2007, first issued by the European Investment Bank, followed a year later by the World Bank. Green bonds started off slowly but accelerated in 2013 when the International Finance Corporation used its first green bond, reaching a value of over US$600 billion in 2024, according to American giant finance company S&P Global. The US is one of the largest sources of green bonds. The market for "sustainability bonds," which combine green and social bonds, has exploded to US$4 trillion today.[111]

Green bonds make money for investors, but do they help the environment? Much evidence says no. Green bonds are widely criticized for allowing companies to utilize the capital raised for projects promoting sustainability that, in fact, have minimal or even negative climate impact. There are many examples.

The operator of China's Three Gorges Dam, widely criticized for environmental damage and water pollution, issued US$840 million in green bonds to be used for backing wind-power projects in Europe. The Chinese government also issued green bonds to finance coal-efficiency projects to burn fossil fuels more effectively. A Madrid-based oil and gas company issued green bonds used to make their still-polluting oil refineries more "efficient."[112]

A year-long investigation by environmental organization Mighty Earth alleges that in 2015, tire manufacturer Michelin used a US$95 million green bond to finance "natural, eco-friendly rubber" production in Indonesia by massive reforestation. Investors were never told that Michelin's local partner, Barito Pacific Group, had deforested thousands of hectares of tropical rainforest and wildlife habitats just prior to the launch of the project and that, after the contract was signed, it deforested thousands of hectares of pristine rainforest to make way for monoculture rubber plantations.[113]

Green finance consultancy Trellis echoes many others in the investment world, calling ESG the "Wild West" of finance. For all the salutary change it may be creating in financial circles, the ESG landscape remains treacherous and unforgiving, their analysts write. Data is uneven and incomplete. Many companies and banks fail to align their public commitment with their prices. "ESG and sustainable finance have long been, and remain, the Wild West, filled with myriad gunslinging parties staking their claim, often duking it out with competitors and critics, without any sense of who's truly in charge — or even who the 'good guys' are."[114]

Voxeurop, an independent online media outlet made up of journalists from over 30 countries, reported in 2024 that banks and asset managers using the ESG Planet Saver funds in Europe were, in fact, investing in some of the world's biggest carbon emitters, deepening the climate crisis. Their reporters explained that the contradiction stems from a loophole in the EU rules that allows green funds to be invested in greenwashing projects, while officially claiming to be for the public good. The top 10 "green" finance operators in Europe, including BlackRock, J.P. Morgan Asset Management and Deutsche Bank Wealth Management, are responsible for more than a quarter of all investments by EU-regulated green funds — about US$87 billion — in the 25 biggest greenhouse gas emitters.[115]

Even the investment industry itself admits that green bonds have failed in their promise. In a 2022 report, S&P Global identified more than 300 funds that use green or environmental language to describe their approach and found that only 12% are on track to meet their Paris Agreement commitments, undermining the eco-friendly or climate-conscious signals they send to investors.[116]

Bloomberg News echoes these findings. Almost all green bonds issued in the US fail to drive real action to tackle climate change, it noted in a 2024 report. Green finance is supposed to be used to fund projects that offer distinct environmental benefits beyond standard practice. A 2024 analysis by the US National Bureau of Economic Research of green bonds sold between 2013 and 2022 found that the vast majority were used for refinancing ordinary debt, and only about 2% of proceeds were used to initiate projects with novel green features. "A cynical interpretation of the results is that the green bond market is largely a financing show," said Pauline Lam and Jeffrey Wurgler with New York University's Stern School of Business.[117]

Carbon trading has been a disaster

Carbon markets are the most visible form of green finance and have been around for decades. American entrepreneur and economist Richard Sandor, called the "father of financial futures," created the Chicago Climate Exchange in 2003 when he was chief economist with the Chicago Board of Trade. He had already developed a market for sulphur-pollution trading to deal with the acid rain crisis in 1991 in the US and used the framework to take the idea of carbon trading to the UN's 1992 Earth Summit. The Kyoto Protocol in 1997 established a formal mechanism for trading in greenhouse gas emissions. Carbon markets are now the largest class of environmental trading markets in the world in terms of both volume and value.

Essentially, offsetting is a carbon trading mechanism that allows corporations to "compensate" for their greenhouse gas emissions — and continue to pollute — by investing in projects that allegedly reduce emissions elsewhere. One UN-issued carbon credit represents an emissions reduction or removal of one metric tonne of greenhouse gases somewhere. When the company invests in a carbon offsetting project, it receives a credit that can then be traded. Until recently, carbon credits have mainly been traded by companies on an unregulated market. At the November 2024 climate COP, new rules allowing wealthy polluting countries to buy carbon offsets from developing nations were created.

Closely linked to carbon offsets is another UN program called Reducing Emissions from Deforestation and Forest Degradation (REDD) that was introduced into UN climate talks in 2005 and incorporated into the Paris Agreement. REDD is a highly financed payment for environmental services scheme that offers financial incentives to reduce deforestation and therefore lower the greenhouse gas emissions resulting from forest loss. Hundreds of millions of acres of lucrative land in the Global South have been acquired by foreign governments and transnational corporations through this project that some call "carbon neocolonialism."

REDD is a major boon for the carbon markets industry as it is funded through the sale of carbon credits, allowing greenhouse gas polluters to claim that the climate damage from their own emissions has been cancelled out by someone foregoing a planned emissions activity elsewhere. But the reality is more complicated, explains the German-based Heinrich Böll Foundation: "Forests are much more than stores of carbon, and the root causes of deforestation are complex and political as well as financial in nature. In addition, forest use is often caught up in conflicts over rights to access and ownership, and carbon storage levels in forests fluctuate naturally. . . . The result 15 years on is that REDD has become a huge accounting exercise, but it has been unable to tackle the root causes of forest loss, with the outcome being the continued destruction of forests around the globe."[118]

Eucalyptus plantations for carbon credits are deeply controversial in Africa and many other parts of the Global South because they are monoculture crops that displace local indigenous plants, trees and crops, damage the soil and guzzle and contaminate local water sources. Eucalyptus trees grow quickly, so they are harvested early, undermining the promise that these are forests that will serve as carbon sinks into the future. Eucalyptus is a key resource for the global pulp industry, and Portugal has become Europe's top producer of eucalyptus-based cellulose, both at home and in Africa. Eucalyptus plantations now cover over one-quarter of Portugal's forest area, but according to the Africa-based Oxpeckers Investigative Environmental Journalism Centre, the popularity of this "lazy crop" has dwindled as the trees are invasive and, when left wild without forest thinning, are a "ticking time bomb" for wildfires.[119]

Mozambique was the first country to receive payments for emissions reduction from the Forest Carbon Partnership Facility, a World Bank–managed fund of governments and businesses that promotes REDD. World Rainforest Movement reports that the eucalyptus plantation company Portucel Moçambique, created by Portuguese giant the Navigator Company, one of Europe's largest pulp-and-paper corporations, obtained a government concession to use 356,000 hectares of land for 50 years to plant eucalyptus trees. The government's own reforestation plan includes a target to increase the plantation area to one million hectares by 2030. Local communities are reporting dried-up streams and wells, harassment from company security guards, displacement from their farms and forests and little of the compensation promised. World Rainforest Movement accuses the company of obscene accumulation.

"One of the statements made by Portucel in its advertising cannot be denied: that its activity means 'wealth and added value generation in the country.' Doubtless, the acquisition of cheap lands in the Global South by corporations from the Global North, with the support of international agencies, associated with the employment of highly exploited cheap labor, represents gigantic possibilities for generating wealth IN the country. It does not mean, however, that the wealth REMAINS in the country, and even less with the PEOPLE of that country. The case of Portucel/Navigator, the self-proclaimed 'world's most sustainable forestry company' is an example of how the propaganda of sustainability and of social benefits legitimizes a process of primitive accumulation that allows a corporation from the Global North to transform people and nature into mere resources of production (labor and land) and insert them — at a very low cost — in the circuit of expanded reproduction of its capital."[120]

Green Earth Africa, a major agricultural landholder in Zimbabwe and Zambia, boasts that the trees it is selling for carbon credits are very cheap — US$1 per tree. It assures investors that the wood of its teak and eucalyptus plantations is a double investment: investors buy it for the credits, but when the forest is logged, the wood is valuable for its strength in construction and as utility poles for electricity lines. "Companies are able to finance the planting of trees or purchase trees in our plantations and benefit not only from the green credentials and carbon credits produced from these trees which Green Earth Africa will award to its company but it will also benefit in seven years' time from the sale of these eucalyptus products and the customer will have the choice to reinvest these proceeds back into new plantations or receive a dividend for the sale of the products. For every number of trees you buy, you will also be awarded the carbon offsets which are created from these plantations," says the company.[121]

Small wonder that Adrienne Buller says that money really can grow on trees.

The carbon market system has been plagued by failures from the beginning. "The Global Carbon Trading System Has Essentially Collapsed," announced *The Guardian* in 2012, having lost "tens of billions" of dollars after the collapse in the price of the UN credits.[122] Yet the program has expanded dramatically since its inclusion in the Paris Agreement, driven by a host of private-sector players not willing to give up on their new frontier.

Heinrich Böll has documented stunning failures, including a 2015/2016 Green Climate Fund US$28.2 million payment to Colombia for alleged greenhouse gas emission reductions when, in fact, deforestation levels rose sharply in 2016, and a 2020 US$103 million payment to Indonesia for reducing emissions several years before, when, in fact, fires in forests and peatlands had released unprecedented greenhouse gases.

The failures are legion. A 2022 investigation by the NGO Carbon Market Watch found that millions of "hot air" carbon credits designed to curb deforestation in Colombia were sold to a Colombian fossil fuel company as a substitute for paying the country's carbon tax, costing the country millions in tax revenue and undermining its climate goal.[123] The UN has called into question the transparency of a 2021 nature conservation agreement between Hoch Standard, a Singaporean company, and the Sabah state of Borneo that gives the company rights to carbon and "other marketable ecosystem services" from more than half of the state's forests. One major concern is that Indigenous Peoples, who make up half the population, were not consulted.[124]

The Indigenous people of the Batéké Plateau, a vast savannah in the Republic of the Congo, are also being displaced to make way for a carbon offsetting tree-planting project of oil giant TotalEnergies. Acacia trees are being planted across 40,000 hectares of land to generate carbon credits for the company. The local Indigenous farmers and their ancestors have lived on the land for as long as any can remember, but they do not have individual title to the land, which is owned by the government. A 2022 investigation by Greenpeace reported that the Congolese government passed a law a year before the operation began, setting aside the land for it, later signing a lease promising to evict local people with ancestral claims to the land. "Their trees over there are for their benefit, not ours. We have no interest in their money," said one farmer. "Since the company arrived here," said another, "we no longer have fields. I feel like these people came to kill us on our own land."[125]

In May 2024, Greenpeace reported that the Shell plant in Alberta registered millions of carbon credits for reductions in greenhouse gas emissions that never happened, allowing the company to turn a profit on its fledgling project. Between 2015 and 2021, it registered 5.7 million tons of "phantom credits," making more than CAN$200 million, selling some of the credits to other oil companies operating in the region. To add insult, the report pointed out that Canadian taxpayers had funded 93% of Shell's project during those years.[126]

Washington-based Food & Water Watch also reports that "ghost credits" are a rampant problem that can actually increase carbon emissions, as they allow corporations to pollute more regardless of the efficacy of credits. The organization points out a distressing irony in the whole debate: one of the promises of carbon trading is permanence — that the emissions from projects such as forest protection and reforestation will be prevented or stored for hundreds, even thousands of years. But many of the carbon offset projects are facing an existential threat from wildfires, exacerbated by climate change. Food & Water Watch mapped forest offset projects in California and overlaid them with data on wildfire hazard potential from the US Forest Service and found that a significant portion of these forest offsets are at risk of literally "going up in smoke."

There are already examples. Over the course of August 2020, a large wildfire burned 72% of Oregon's largest forest offset project involving Microsoft and other corporations. The trees that burned in this fire stored the equivalent of the greenhouse gases produced through the course of a year by more than 700,000 cars. If the California forests set aside for carbon offsets that are currently on high alert for wildfires were to burn, they could release the equivalent to the yearly emissions of over 11 million cars. "It is clear that forest carbon offsets and other pay-to-pollute schemes are climate scams that allow corporations to increase carbon emissions at the expense of the environment and environmental justice communities. Instead of market-based schemes, we must address the root of the problem by transitioning away from fossil fuels and banning factory farms, both of which contribute significantly to deforestation and climate change."[127]

A particularly contentious carbon market project is one by Blue Carbon, a company owned by the UAE royal family that has secured rights to more than 100 million hectares of land in five African countries to generate carbon credits for the UAE. In October 2023, Zimbabwe signed away control over 20% of the country's land in the deal, and Blue Carbon will receive 70% of the revenues from selling the carbon credits. It also signed a deal for millions of hectares of forests in Kenya. So far, the company has made deals in Africa amounting to a total land area the size of the UK, and the UAE has pledged to buy US$450 million of African credits by 2030.[128] This is a petrostate that says it plans to extract every last barrel of its oil 50 years from now, when its reserves are projected to dry up.

In 2023, an in-depth investigation by Corporate Accountability — a non-profit corporate watchdog — and *The Guardian* found that the majority of offset projects that have sold the most carbon credits are "likely junk," and have such fundamental flaws, they cannot be relied upon at all to cut planet-heating emissions. Researchers looked at data from AlliedOffsets, the most comprehensive emissions trading database in the world, and found that 39 of the top 50 emissions-offset projects were worthless, and eight others were "problematic." These include a giant forest conservation project in Zimbabwe that had so many exaggerated claims, it was described by *Bloomberg* as "having more financial holes than Swiss cheese." This project's emissions cuts were overestimated by 5 to 30-fold.[129]

A 2024 investigation of Australia's domestic carbon offset market found it to be a "failure on a global scale" doing little if anything to help address the climate crisis. The academics analyzed 182 projects in the country's arid and semi-arid desert areas and found that forest cover had either barely grown or gone backwards in nearly 80% of cases. Australia's forest-regeneration project is the world's fifth biggest nature-based offsets project and more than 37 million carbon credits worth almost US$1 billion have been issued.

Megan Evans, with the University of New South Wales and co-author of the study, said: "What this means is that the projects are not actually sequestering the amount of carbon claimed, and we've got a whole bunch of carbon credits in the system that don't represent one tonne of CO_2. Most of these credits are being used to offset heavy emitters under the safeguard mechanism, so we're not actually reducing carbon emissions at all. The overall outcome is we're increasing the amount of carbon pollution. We're ultimately getting worse outcomes for the climate than if we didn't have these [forest regeneration] projects."[130]

A damning report by the global climate science institute Climate Analytics underlines the threat to the climate when Australian companies invest in carbon offsets such as tree planting and forest regeneration in other countries. These offsets are, by design, not permanent, meaning much of the carbon sequestered is lost to the atmosphere. In fact, said the scientists, they calculated that for every Australian carbon credit unit generated to offset one tonne of CO_2-equivalent emissions from LNG production in Australia, about 8.4 tonnes of CO_2 are emitted globally, and for every tonne of emissions from Australian coal mines, up to 69 tonnes of CO_2 emissions are released elsewhere. In promoting carbon markets for its fossil fuel industry, the Australian government is allowing them to continue to pollute at home and do "massive" climate damage abroad.[131]

A 2023 *Euronews* investigative report tallied the main reasons for the failure of carbon trading: artificial inflation of baseline emissions to take credit for what the project did not do; reliance on vague and often inaccurate predictions, including promises of carbon sequestration that may take 50 years to realize; forceful eviction of local communities and Indigenous populations; high risks to planted forests from logging and cattle grazing after the project ends; and, most importantly, failure to contribute to achieving additional overall climate benefits, compared to if the project had not existed.[132]

Yet carbon markets thrive

That carbon gravy train keeps making profits for corporations. In 2024, 43% of Tesla's net income came from selling carbon credits to other fossil fuel carmakers that had failed to produce sustainable vehicles in line with US and EU environmental goals, earning US$2.8 billion for Elon Musk. Because Tesla sells its credits to other automakers, even consumers who do not want to support Musk and his politics by buying his cars can still be indirectly lining his pockets, reported investigative journal *Follow the Money* in 2025. Tesla made more than US$9.4 billion on carbon offsets between 2020 and 2025.[133]

In October 2024, 60 renowned scientists from around the world warned that carbon offsets used by corporations are not cutting emissions overall and, in fact, are hindering the energy transition. They called on the world to take what they call the "real zero" pledge in place of "net zero," which they call a "counting game."

"A reliance on carbon offsets without the needed emissions reductions is dangerous and detrimental," said Katrin Meissner, director of the Climate Change Research Centre at the University of New South Wales. "To keep global warming within the guardrails of the Paris agreement, the timeframe is now so tight that there is no space for companies to use offsetting to continue high-carbon activities. We need to turn the fossil fuel taps off, all of them."[134]

Not only does the carbon market continue to grow, so does Big Oil. After posting record profits in 2022, the big energy companies had another banner year in 2023, the five biggest (Shell, BP, Chevron, ExxonMobil, and TotalEnergies) paying their shareholders an unprecedented US$111 billion. The Union of Concerned Scientists notes that these record profits came as people around the world suffered billions in damage from climate disasters — damage that governments and communities had to clean up.

The union quotes the CEO of Chevron bragging about his company's record profits and rising production levels: "In 2023, we returned more cash to shareholders and produced more oil and natural gas than any other year in the company's history." The scientists say this comment shows "an atrocious disregard for the fossil fuel industry's harmful impacts on the world and for global efforts to confront climate change and prioritize human rights."[135]

A March 2025 report released by NGO the Fossil Fuel Non-Proliferation Treaty Initiative details that Australia — a party to the Paris climate agreement — has quadrupled exports of emissions-heavy fuel sources since 1990, leading initiative president Kumi Naidoo to call the country a petrostate. The report finds Australia exports 48% of the world's traded metallurgical coal, 19% of the world's thermal coal and 20% of the world's liquefied natural gas. Overall, Naidoo told *The Canberra Times*, only Russia and perhaps Saudi Arabia send a larger quantity of fossil fuels offshore.[136]

According to the Natural Resources Defense Council, rather than using these profits to invest in their renewable energy portfolios, Big Oil is expanding their fossil fuel development, "putting shareholder profit over the planet and people's lives."[137] And their lobbying against government regulations has not abated even as they claim to promote a nature-friendly image. Greenpeace reports that in 2024, Big Oil spent more money fighting California's proposed climate policies than in any previous year.[138]

And they are doing it using taxpayer dollars to fund their exploration and development. Not only is the public paying for the development of new oil and gas wells, it is also paying for the cleanup of unplugged wells. Globally, government subsidies for oil, gas and coal are at record levels, having reached US$7 trillion in 2022, reported the International Monetary Fund.[139]

Federal and provincial governments in Canada are subsidizing the Alberta tar sands — the world's largest, and according to *National Geographic*, dirtiest, fossil fuel project — by at least US$13 billion a year reported Canadian NGO Environmental Defence in 2024.[140] Overall Canadian subsidies to the oil and gas sector, including financing for the Trans Mountain pipeline project, topped US$20 billion in 2024, reported *Yahoo Finance*.[141] In March 2025, *Canada's National Observer* reported that 180 oil and gas companies plan to spend more than US$425 billion in developing new oil and gas fields in the next decade, and annual investments in Canadian energy production are expected to reach the highest levels ever recorded in

Canada by 2030.[142] Meanwhile greenhouse gas emissions continue to grow. A January 2024 report published in *Science* found that oil companies in the tar sands are emitting up to 6,300% more pollution than they have been reporting.[143]

During the spring 2024 UN Permanent Forum on Indigenous Issues, the Indigenous Environmental Network (IEN) said that 25 years of evidence was more than enough and called for a permanent end to carbon markets. IEN's executive director, Tom Goldtooth, told the assembly, "We are long overdue for a moratorium on false climate solutions like carbon markets. It's a life and death situation with our people relating to the mitigation solutions that are being negotiated. Article 6 is a smokescreen, a loophole that keeps fossil fuel polluters from agreeing to phase out carbon."[144]

Next in line: biodiversity offsets and trading

While it has been used in limited ways for decades in a number of countries, biodiversity offsetting, also called "nature markets" and "conservation offsets," is a more recent manifestation at the international level than carbon offsetting and has consequently flown under the public radar. Biodiversity trading places a value on a habitat, plant or animal, and a credit is created that can be offset, or traded. This allows a company or government to continue to harm a local habitat while claiming to conserve natural assets elsewhere.

Indian conservation news website *Mongabay* explains biodiversity financing as the practice of raising capital and managing funds for biodiversity conservation. It says that it is still mostly dependent on funds from governments, international development banks and private philanthropy. But the industry focus is shifting toward sourcing finances across a variety of public and private funders through loans, grants, tax incentives and market mechanisms, which include tourism-related taxes and fees, debt-for-nature swaps, conservation trust funds and direct payments for environmental services.[145]

Biodiversity offsetting is the latest chapter in a long history of debates that has pitted the environment against development, says *Carbon Brief*, a UK-based website covering climate science and policy. To counter strong proposed nation-state regulations to protect biodiversity and wildlife, mining and energy industry consortiums teamed up with conservation organizations such as The Nature Conservancy to instead allow industry to introduce voluntary land-use commitments and offsetting. It was only a matter of time before biodiversity offsetting would become a formal part of the Convention on Biological Diversity process.

Carbon Brief says that while biodiversity-offset markets have so far escaped the intense critique that carbon offsetting has been subjected to, they are growing in prominence and were included as one of the ways to finance a global deal for nature approved at the Biodiversity COP15.

At the October 2024 Biodiversity COP in Colombia, a framework laid out principles and guidelines to develop "high integrity" biodiversity credits by the International Advisory Panel on Biodiversity Credits, a powerful new global network dedicated to promoting biodiversity trading. "High integrity biodiversity credits represent a unique opportunity to scale up private finance for nature and engage the private sector in the ecological transition," said co-chair Sylvie Goulard at an event organized by the European Commission.[146]

Like carbon offsets, explains *Carbon Brief*, biodiversity offsets are built around the idea of "no net loss" — the assumption that damage to ecosystems wrought by large-scale development projects can be balanced or outweighed by preserving or "producing" nature elsewhere. Unlike carbon offsets, there is currently no international marketplace for buying or selling biodiversity credits or offsets, although there is political will in France and the UK for such a market.[147] In December 2023, French President Emmanuel Macron announced his intention to create an international carbon and biodiversity exchange.

There are many governmental, corporate and foundational institutions promoting nature markets. The Intergovernmental Science-Policy Platform on Biodiversity and Ecosystem Services is an influential body established in 2012 by 94 governments to strengthen scientific cooperation for biodiversity protection. Often referred to as the IPCC for Biodiversity, it is housed at the UN Environment Programme. In December 2024, the platform published a report "revealing huge economic and business opportunities for accelerating biodiversity action" and predicted that immediate action to address the biodiversity crisis "could unlock massive business and innovation opportunities, generating US$10 trillion" by 2030.[148]

NatureFinance is a Geneva-based organization whose mandate is "integrating natural capital into the world's sovereign debt markets, including scaling the issuance of sustainability performance-linked sovereign bonds" and "advancing effective governance of nature markets (those markets that explicitly value and trade nature) including voluntary carbon credits, conservation, soft commodities."[149] NatureFinance launched a taskforce on nature markets that submitted its final report at the August 2023 Amazon Summit for a Fair Nature Economy in Belém, Brazil. "'Markets' — including voluntary carbon credits, conservation and soft commodities — produce and trade over US$7 trillion worth of goods and services annually, equivalent to the 3rd largest economy after the USA and China. A new era of natural markets is emerging," said the report.[150]

The Finance for Biodiversity Foundation was established in 2020 to promote financial investments in biodiversity. As of 2025, 200 financial institutions representing 28 countries and over US$23 trillion in assets have joined and taken the Finance for Biodiversity Pledge, calling on global leaders to commit to protecting and restoring biodiversity through their finance activities and investments. The signatories are aiming to have influence at the UN biodiversity COPs, and members include major banks, portfolio and asset management companies, and pension and investment funds. HSBC Global Asset Management CEO Nicolas Moreau stated in 2020, "We are committed to putting natural capital on the map as a new asset class that can no longer be overlooked by investors."[151]

And in February 2025, Goldman Sachs Asset Management launched its first biodiversity-focused bond fund, "offering fixed-income investors a way to support nature conservation." The fund aims to reach US$500 million in assets over the next three years.[152]

"Natural asset companies" eyeing national parks around the world

A major target of biodiversity investing is in underfunded national parks and conservation areas around the world, what the UN calls "protected areas." The protection and restoration of wild spaces is a key international commitment. Governments around the world, as well as the UN and the World Bank, have jumped on this bandwagon with enthusiasm, with global initiatives such as 30 x 30, whereby governments pledge to designate 30% of their land and oceans as protected areas by 2030. Launched in 2020 by the UN High Ambition Coalition for Nature and People, its stated purpose is to expand nature conservation and biodiversity protection to mitigate climate change. Member countries committed US$5 billion to the project, and as of 2025, 122 countries have taken the pledge, many of them with very good intentions.[153]

It's a daunting order. The UN Environment Programme 2024 report on the 30 x 30 project noted that some headway has been made, but not enough. A decade ago, about 14% of global land and inland waters and 5% of oceans and coastal areas were documented as "protected and conserved areas." By 2024, these numbers rose to 17.6% and 8.4% respectively, a measure that shows "some headway," but not nearly enough, said the UN. The press release notes, "This leaves a land area roughly the size of Brazil and Australia combined, and at sea an area larger than the Indian Ocean, to be designated by 2030 to meet the global target. Over the next six years, the global network will need to be urgently expanded by a further 12.4% inland and 21.6% in the ocean."[154]

Where can nations find the funding? Managing national parks and conservation areas is expensive for all governments, especially for many in the Global South. There is an increasing trend within the management of parks to form relationships with the private sector, reported *Mongabay* in 2016. Public-private partnerships use private money to fund national parks and other protected areas, in many cases, commercializing them to pay off the vital services needed for maintenance services. While the government still owns the actual land, a private consortium runs it.

South Africa–based NGO Africa Parks takes full control of entire parks in several African countries and runs them top to bottom, keeping all the money it makes, giving none to the governments. In one of Colombia's most pristine reserves, Tayrona National Natural Park, reported *Mongabay*, 90% of the land is held by private individuals, and the company operating the tourist concessions has been accused of land grabbing and controversial development within the park.[155]

Robert Williams, a law professor and faculty chair of the Indigenous Peoples Law and Policy program at the University of Arizona, warns that the 30 x 30 project is being used to displace Indigenous Peoples in many communities. In the name of rewilding nature and creating "protected areas," millions are at risk. Williams says that the project is sold as a way to protect nature, but it opens opportunities for economic development, such as ecotourism and carbon credits. He points out that people have lived on these lands and cared for them for millennia.

"Indeed, an estimated 476 million Indigenous People dwell on lands that are home to 80 percent of the world's biodiversity. When governments decide that nature conservation and potential revenue from it take priority over existing human activities, too often they resort to eviction, destruction of agricultural fields and confiscation of livestock, sometimes through stupefying violence, to get residents off the land." He cites tiger preserves in India to attract foreign tourists that have displaced Adivasi people and the planned removal of nearly 100,000 Maasai people from Tanzania's Ngorongoro Conservation Area to increase the number of tourists who come to see the annual Great Migration.[156]

Scientific American reported that in the Congo Basin, armed eco-guards brutally evicted Indigenous Pygmies from the rainforest to carve out protected areas following a 2010 resolution by the UN Convention on Biological Diversity to dedicate 17% of Earth's terrestrial surface to nature by 2020. "Yet the protected areas are surrounded by or sometimes even overlaid with oil, mining or logging concessions. Unsurprisingly, chimpanzee, gorilla and elephant populations have continued to decline even as Pygmy peoples have been consigned to poverty and misery."[157]

A fall 2024 report by the Oakland Institute exposed abuse of Indigenous Peoples in the eastern Democratic Republic of the Congo (DRC) in the name of conservation. "Removing Indigenous communities from lands earmarked as a protected area has created a political vacuum filled by outside commercial actors seeking to exploit the DRC's natural resources. This conservation model negatively harms both biodiversity and people, while contributing to the ongoing political instability in the region," said the organization's policy director, Frederic Mousseau. The press release also notes, "Uganda and Rwanda, who receive every year millions of dollars of military and economic assistance from Western countries, are profiteering from the exploitation of the DRC's vast gold, tantalum, and cobalt reserves, while fueling the conflict in the already war-ridden country."[158]

Survival International calls the use of biodiversity and carbon offsetting in protected areas "blood carbon" and says they are part of a new move to commodify nature. "Protected Areas are still being pushed by big conservation NGOs (like WWF and WCS) as the 'solution' to our very real environmental problems. . . . These projects put a price on nature, treating Indigenous and local communities' lands as a carbon stock to be exchanged in the market so polluters can keep polluting, the conservation industry can get its hands on billions of dollars, and speculators can profit. This leaves Indigenous peoples and local communities dispossessed and stripped of their livelihoods. 'Nature-Based Solutions' offset schemes are carbon colonialism and won't stop the climate crisis. Survival is campaigning to end carbon projects in Protected Areas where the rights of Indigenous peoples are violated. . . . Carbon colonialism is killing people and the planet."[159]

Worryingly, a 2021 World Bank blog promotes the use of national public parks — protected areas — around the world to meet the 30 x 30 pledge. While public parks do, of course, count in the 30 x 30 goal, the blog authors have something else in mind. They recommend shifting the ownership of these parks from governments to a public-private partnership model, where the government still technically owns the land, but its use is privatized. To pay for the pledge, the Bank would turn to public pension funds so large they now, in effect, "own a slice of the global economy," and public-companies, often referred to as "sovereign wealth funds," that governments invest in on behalf of their citizens.

As of 2021, the global top 10 public pension funds and the global top 10 sovereign wealth funds controlled US$15 trillion in assets. Together, they could fund Natural Asset Companies (NACs) that would invest in the newly privatized parks. The authors explain,

> The land would still be owned by the national government with universal investors (international and local pension funds) making an equity investment in a NAC, which would have the sole purpose of maximizing nature and biodiversity. Sustainable revenue generated by the parks (through concessions for tourism, agriculture, sustainable timber or new revenue streams — including the sale of carbon or biodiversity offsets) would create "dividends" that

> could contribute to sustainable livelihoods for local community members, be reinvested in the NAC, shared with the government and invested in new PAs.
>
> The NAC would be valued based on the intrinsic value of the natural asset as well as the value of the revenue streams coming from ecosystem services which can be monetized (now and in the future). . . . Ultimately, the NAC could be listed on a stock exchange (preferably a local emerging market exchange to support capital market development) enabling all interested investors to invest in nature. This would create a new and uncorrelated asset class.[160]

Note that this model would take public funds, public parks and protected areas funded by the public and governments and protected by government regulations and hand them over to the service of private capital and the control of the market.

It is not just in the Global South or Indigenous communities of the Global North that such abuses take place. Billions of people who live in urban centres around the world depend on state parks and protected areas to experience nature. If these protected areas are sold to private interests in the name of conservation, billions could lose their only access to natural ecosystems, forests, lakes and rivers.

In fall 2021, the New York Stock Exchange (NYSE) partnered with Intrinsic Exchange Group to open up investment opportunities in "nature's economy" by creating a "new kind of listing vehicle" — natural asset companies. *Fortune* explains, "Using NACs, governments, farmers, and other owners of natural assets will be able to form a specialized corporation that holds the rights to the ecosystem services produced on a given chunk of land, services like carbon sequestration or clean water. Then the company will tap the U.S. public markets by way of the NYSE like any other entity would. . . . In return, investors will get access to a new form of sustainable development — a space that has enthralled the likes of BlackRock CEO Larry Fink over the past several years."[161]

Big energy, mining and logging companies were excited about what would have been the largest attack on public US lands and water in modern US history. Initially, in its summer 2025 "big, beautiful" bill, the Trump administration proposed to lease 2 million acres of public lands — an area larger than Texas and California combined — to the private sector. "Expanding oil and gas leasing on public lands and waters through mandated lease sales would turn parks, coasts, wildlife refuges, and sacred landscapes into drilling and mining zones," said the non-partisan policy institute Center for American Progress.[162] Strong opposition from many quarters, including the outdoors recreation industry, caused Senate Republicans to remove this provision from the bill. But Americans who care about their public parks and wildlife are on notice that the government might try again at some future time.

The UK has launched a biodiversity unit market and "habitat banks" — areas of land that are "restored to create a bank of offset units that can then be allocated and sold," according to Wild Capital, a British "elite consultancy" company. Wild Capital is clear about who its clients are: "Land Promoters, Developers, Brokers and Affiliates." One product they offer: "Bespoke schemes that deliver multiple habitat types, including high-distinctiveness habitats, specifically delivered for individual developments or pipelines. Developers have access to the specific number and type of offsets they need quickly and at competitive prices within the location where they need them."[163]

In Scotland, a new breed of buyers is snapping up old estates, creating deep controversy in local communities. In February 2025, *The Washington Post* reported "green lairds" — energy companies, fast-fashion billionaires and private equity funds with the support of governments — are planting forests and restoring peatlands to serve as carbon sinks. The end goal is biodiversity restoration in exchange for carbon credits, either to allow the investors themselves to continue to pollute or to sell shares and credits to other polluting companies. Some want to use the rewilding of the land to show a profit by selling rewilding shares and credits. Many involve a "side hustle" of luxe tourism.

Scottish newspapers are filled, noted the *Post*, with tales of failed forest planting and improbable plans to bring back long-gone animals, alongside articles about soaring land and home prices local families cannot afford. Alastair McIntosh, a land reform scholar, fears that the "carbon rush" could be a fresh driver of upheaval, disempowerment and depopulation: "People are wary of the big corporations coming in, buying massive tracts of land, way over the asking price, to plant monocultures of spruce for carbon credits," said one local farmer. "We're trying very hard to keep the town alive," said Martin Mellor, a local activist in the town of Tayvallich. He is not opposed to rewilding but questions it being driven by carbon and biodiversity credits. "If all it does is just let the baddies carry on being bad, then I don't know if that is a good thing for the planet, or even a decent business model."[164]

New South Wales in Australia launched its own biodiversity offset scheme. It allows landowners to establish a conservation deal on their property that generates credits that then can be sold on a biodiversity market to compensate for environmental destruction elsewhere. The spot price for squirrel glider credits in late 2022 was AU$426, and koala credits were going for AU$600. A review that year by the state's auditor general revealed serious flaws and conflict of interest concerns, leading to further investigations. Jacqui Mumford, chief executive of the Nature Conservation Council of NSW, said the review was "utterly damning," reported *The Guardian* in 2022, and has failed by every measure. She said that offset schemes lock in existing environmental decline and reduce nature to a "bunch of financial formulas" that do not capture the true value of unique and rapidly disappearing bushland.[165]

Undeterred, in 2024, Australia unveiled its Nature Repair Market, a voluntary project whereby biodiversity credits earned by farmers and landowners who implement projects to restore and protect the environment can be sold to private buyers, including corporations and asset managers wanting to claim environmental points. Penny van Oosterzee, with James Cook University, has concerns about the project. Noting that Australia is a world leader in species extinction and environmental decline, she warns that biodiversity markets, if not carefully governed, are as likely to fail as carbon markets and calls for the government to speed up the progress of proposed environmental law reform.[166]

Canada is also preparing biodiversity offsetting legislation, raising alarm bells for Canadian environmental organizations. Greenpeace Canada says the draft legislation "flies in the face" of the constructive role Canada has played in international biodiversity meetings, choosing now to follow a "misguided approach" that will only exacerbate the biodiversity crisis. Greenpeace is particularly concerned with plans for "habitat banking," as in the UK: "This opens up the possibility of a market-based system wherein wealthy developers or corporations simply pay a fee in order to destroy biodiversity — a pay-to-slay mechanism.

"Ecosystems and biodiversity can take centuries, or even millennia, to recover even in the best circumstances. If governed by a for-profit corporation, how do we ensure the kind of necessary long-term implementation to ensure restoration actually occurs? And will climate change or other weather events disrupt restoration potential unexpectedly in the future? How will this policy affect old growth forests? It is not clear whether it could lead to situations in which an old growth forest is logged, but this logging is deemed 'net positive for biodiversity' through a subsequent tree plantation. . . . This trash now / pay later mentality is a big part of what has gotten our society into the current crisis. Getting us out of it requires shifting our attitudes and policies away from seeing nature as a cheap commodity that can be moved around at our convenience."[167]

In spite of such concerns, investors see Canada, with its biodiversity wealth, as a new and lucrative frontier. In February 2025, the

Ottawa-based Nature Investment Hub released a survey claiming that investors are "keen" to grow their nature conservation investments in Canada. The 60 respondents, including investment firms, pensions funds, foundations and banks, represent over US$180 billion in assets.[168]

Physicist and environmentalist Vandana Shiva says that natural asset companies are the next step in the commodification of nature for profit and will use the debt crisis as a takeover of the real resources and real wealth of the Global South — its forests, lands, rivers and biodiversity. Local communities will serve as agents of the new "owners," the banks, foreign corporations and investment companies.[169]

Valuing biodiversity is extremely difficult because of how complex and unknowable any ecosystem is. Many schemes simply divide the ecosystem into units of species or plants. The Conservation Foundation, an American land and watershed conservation organization operating since 1972, describes one of their biodiversity credit pricing schemes as "a unit of space and time: a set number of square meters being actively conserved for a defined number of years. The credits enable companies to buy units of conservation projects simply as an investment in the cost of doing business."[170]

Biodiversity and land theft also entail seeds and patents. La Via Campesina, an international movement representing millions of peasants, landless workers, Indigenous Peoples, small farmers and fishers, has long been concerned that the system to defend peasants' rights to their seeds and genetic resources set up decades ago by the UN Food and Agriculture Organization has been violated by big seed companies protected by rich countries. The organization says that the establishment of a system to gather information on the majority of food we derive from plants into a global pool of genetic resources has been a boon to the big seed companies, who do not recognize the rights of peasants and small farmers or compensate them.

A new concern is the technology that allows open access to genetic sequencing digitally, digital sequence information, touted as a nature-friendly development but seen by those who are daily keepers of the land as another way to steal their seeds and knowledge. Presenting at 2022's Biodiversity COP15, the International Planning Committee for Food Sovereignty, representing millions of small-scale food producers, was very clear. "Some propose Digital Sequence Information to save biodiversity, as if you can just dematerialize our Mother and piece her back together and hope she functions better. Turning nature into capital is anything but 'living in harmony with nature.' The 'nature-based solutions' debated here and at the climate COP put nature on a ledger and then sell her to polluters at the expense of biodiversity, land, and the rights of Indigenous Peoples, small-scale food producers, and local communities."[171]

Frederic Hache spent 12 years as an investment banker and understands the market intimately. He is the co-founder of Green Finance Observatory. Hache writes that the beauty of the financialization of nature model for investors is that governments take all the risk and the private sector gets the profits. It favours the corporations and investors in that it removes existing environmental regulations or prevents new, tighter ones, enabling destructive companies to keep their profits. And it creates a new asset class for financial institutions, a "new playground that is lowly regulated and not very competitive, hence with very fat margins."

Comparing carbon and biodiversity offsets, Hache explains,

> For biodiversity, the problem is the same, but infinitely worse, in the sense that this type of market involves transforming into financial instruments not six greenhouse gases, but millions of species, interdependent on each other. It is simply not possible to standardize and boil down biodiversity to a few tradable financial instruments.
>
> Furthermore, ecological priorities are not those of the markets. These may favour a type of ecosystem that is quicker and less costly to restore, and

> underfunds the protection of species and ecosystem functions critical to our survival.
>
> Academic research has amply demonstrated that conservation policies based on economic incentives are much more fragile than those based on regulations, because destroying is often more profitable than preserving.[172]

Hundreds of organizations signed a 2024 Civil Society Statement asking world leaders to reject this false solution. "International biodiversity markets allow elites of the Global North to continue destroying ecosystems, buying cheap credits from the Global South. Biodiversity offsetting can create conflicts over tenure rights and the use of lands, fisheries and forests competes with agroecology and smallholder agriculture, undermining food sovereignty. It will likely drive land grabbing, community displacements, increasing land inequality, and human rights abuses," says the statement. Indigenous Peoples and local communities — the guardians of the planet's biodiversity — receive only a fraction of the proceeds of offset projects, whereas project developers and financial intermediaries receive the biggest share.[173]

Biodiversity offsetting outsources nature protection to the market, relieves governments of environmental responsibility, allows corporations to continue to destroy the natural world and perpetuates colonial inequities, allowing countries and corporations of the Global North to offshore their biodiversity conservation to cheap areas of the Global South.

Plastics offsets and credits

The proliferation of plastics is a major global environmental and health issue and a direct threat to nature everywhere. Every day, says the UN, we dump the equivalent of 2,000 garbage trucks full of plastics in the world's oceans, lakes and rivers.[174] Plastic contaminants have been found in Arctic ice, the stomachs of ocean whales and in freshwater fish of the Great Lakes, some at over 900 particles per fish.[175] The world's oceans are polluted by a "plastic smog" made up of around 171 trillion particles.[176] Environmental organization Oceana reports that the plastic packaging of products bought online from Amazon alone could encircle the globe 500 times every year, most of which ends up in the ocean.[177]

Fossil fuels are the raw ingredient for plastics; if the plastics industry were a country, it would be the fifth-largest emitter of greenhouse gases in the world.[178] Plastic production involves the use of over 16,000 chemicals, including benzene, a known cancer-causing carcinogen.[179]

Carbon Brief reports that plastics cause triple the greenhouse gas emissions of aviation and are set to drive up oil demand over the coming decade. Petrochemical feedstocks derived from fossil fuels are widely seen as the "last growth market" for global oil demand, says the UK-based website, and plastics could use up half the remaining carbon budget for 1.5°C by 2050.[180] Plastics production is on track to triple by 2050 without an international agreement to slow it.

The bottled-water industry is a major culprit, according to a 2023 report by the UN University Institute for Water, Environment and Health. More than 1 million bottles of water are sold every minute around the world, and the industry shows no signs of slowing down. Researchers report that the bottled-water industry saw a 73% growth in the decade between 2010 and 2020 and is one of the fastest-growing industries in the world. They estimate it will reach sales of US$500 billion by 2030. Most of the 600 billion plastic water bottles in 2021 were not recycled and ended up in landfills and waterways. The waste pile is so gargantuan that it would be enough to fill a line of 40-ton trucks stretching from New York to Bangkok every year.[181]

A March 2025 study by Oceana reported that by 2030, Coca-Cola alone will account for an estimated 1.33 billion pounds (602,000 metric tonnes) of plastic waste entering the world's oceans and waterways every year. That's enough plastic to fill the stomachs of 18 million whales.[182] American non-profit Pacific Institute estimates that the total amount of energy required for every plastic water bottle is equivalent to filling it one-quarter full with oil. Very few plastic bottles get recycled globally, and those that don't take one thousand years to biodegrade.[183]

As governments, communities and international institutions try to tackle the plastics crisis, fossil fuel, chemical and bottled-water corporations are promoting a plastics offsets credit scheme similar to carbon markets, which allows them to continue to produce plastic and grow their industry while offsetting their environmental footprint elsewhere.

Companies pay for a weight of plastic to be collected somewhere in the world, generating a credit that justifies their production or use of the equivalent amount of plastic. The exchange is facilitated by accreditors, explains a report in *Canada's National Observer*, who trade in credits. When companies buy enough plastic credits to offset their own plastic footprint, they can claim plastic neutrality or "net-zero plastic."[184]

Companies such as Vancouver's Plastic Bank, "a for-profit social enterprise company," New York–based rePurpose, "the world's leading plastic action platform" and Australia-based Plastic Collective that partners with the World Bank to "channel private capital to projects that reduce plastic waste" claim to be helping not only the environment but poverty in the Global South by paying low-wage workers there to recycle plastic that would cost far more to recycle in the Global North. The World Bank says there were 160 plastic crediting projects registered globally as of the beginning of 2024.[185]

In a May 2024 exposé, journalist Kate Aronoff explains that most projects involve funding waste-processing facilities and waste-collection efforts in the Global South and are on the cusp of major expansion. The first plastics credit market, 3R Initiative, was only launched in 2021. Founders include Verra — the world's biggest issuer of carbon credits — as well as some of the planet's largest users of plastics, including Danone, Tetra Pak and Nestlé and plastics recycler Veolia, the world's oldest and largest private water company. Another exchange — Singapore-based Plastic Credit Exchange — has already sold millions of dollars' worth of credits, including to the Filipino subsidiaries of Nestlé, Colgate-Palmolive and PepsiCo.[186] When Canadian magazine *The Walrus* asked Plastic Bank's CEO, who uses blockchain technology to keep records of waste collection, if plastic credits are a license to pollute, he said, "You know what? Kinda."[187]

Greenpeace unearthed serious early failures on one of the first plastics offsets projects, a waste-processing facility in a town in west Bali, Indonesia. The scheme was sold as "life changing" for the community, and people would be "trained" to sort the mountains of imported plastics, which would be picked up through a waste collection service and taken to a new recycling factory. But a Greenpeace inspection uncovered a very different reality. Trash pickup had stopped, so families and workers were burning the recycled plastics in their backyards. The new recycling facility was swamped with the overflow from the municipal landfill site and was now bigger than when the project was launched. It also frequently catches fire, emitting foul odours and pollutants. Greenpeace also reported that credits had been issued for plastics that ended up incinerated in cement kilns, basically trading one form of pollution for another.[188]

The Alliance to End Plastic Waste is a Singapore-based industry organization made up of nearly 50 major global companies, including Chevron Phillips Chemical, ExxonMobil, Mitsubishi Chemical Holdings, Shell, Dow Chemical, Suez and BASF, the world's largest chemical producer. Its mission is "to end plastic waste on the environment and protect the planet." Planet Tracker, a think tank that promotes sustainable finance, applauds the goal but says it "has doubts" about the true purpose of the alliance. In a hard-hitting 2022 report, Planet Tracker revealed that the majority of the corporate members of the alliance are also members of the American Chemistry Council, who lobbied hard against establishing a negotiating framework for a global plastics treaty, and both organizations promote a plastics pollution solution that focuses on recycling and recovery rather than the supply of plastics. It points out that eight of the top 20 single-use plastic waste-makers are members of the alliance.[189]

In 2019, the Alliance promised to divert 15 million tonnes of plastic waste from the environment by 2023 by creating a circular economy and recycling through credit offsets. However in a 2024 exposé, Greenpeace reported that five of the major companies of the Alliance had produced 1,000 times more new plastic in those five years than the waste they and all the other member companies diverted from the environment. Said Will McCallum, co-executive director of Greenpeace UK, "The recycling schemes they're promoting can barely make a dent in all the plastic these companies are pumping out. They're letting the running tap flood the house while trying to scoop up the water with a teaspoon. The only solution is to cut the amount of plastic in the first place."[190]

Despite the terrible record of carbon markets, the World Bank and world leaders are open to an expansion of offset markets to the plastics industry. And new "investments" in plastic waste are proliferating. In June 2024, the Coca-Cola Foundation gave US$10 million to the International Union for Conservation of Nature to "enhance" the plastic waste recycling operations on seven African Islands, including Mauritius and Madagascar. In the announcement of the project, plastics collectors are referred to as "local water actors" who will be "empowered" to convert their "local knowledge and lived experiences" into "sector-specific and locally appropriate innovations" to deal with the increase in plastic that is landing on their shores. The foundation says that this deal will spur economic growth by leveraging millions in capital investment "towards enterprise development in the sector."[191]

The fossil fuel and plastics industries are working hard to influence UN negotiations for a global treaty to curb plastics pollution, and creating a plastics market is an ideal business-friendly solution for them. While environmentalists and affected communities of the South demand a treaty that would cover the entire lifecycle of plastic from production to disposal, many in the industry want to limit further regulations on plastics such as production caps or chemical phase-outs.

At the April 2024 plastics treaty negotiations in Ottawa, the Center for International Environmental Law (CIEL) and a number of other NGOs reported that the UN handed out 196 passes to fossil fuel and chemical industry lobbyists such as ExxonMobil and Dow — their delegates outnumbering the combined diplomatic representatives of the European Union. CIEL's Delphine Lévi Alvarès said, "If we end up with a treaty that lets the plastics lobby continue business as usual, it will be because of a failure to safeguard the negotiations from their influence. UNEP and the Intergovernmental Negotiating Committee Secretariat's inaction has created the conditions for corporate influence to further tip an already inequitable representation."[192]

Professor Bethanie Carney Almroth, an ecotoxicologist with the University of Gothenburg, told *The Guardian* that she felt "harassed and intimidated" at the Ottawa conference by employees of a large American chemicals company who formed a ring around her, accusing her of "misrepresenting reality." Another encounter was with a plastic packaging representative who barged into the room where she was speaking and shouted at her. Carney Almroth said she has had to take security measures to avoid surveillance at the meetings as people would walk behind her to try to film what was on her screens: "I've been harassed and intimidated lots of other times, in lots of other contexts, at off-site meetings, at side events, also at scientific conferences, via email and so on."[193]

The powerful lobbyists got their way at what were to be the last of the planned UN conferences on a plastics treaty held in

Busan, South Korea, in December 2024 and Geneva, Switzerland, in August 2025. The countries again failed to reach a consensus on the scope of the treaty, with a small number of oil-producing nations, including Saudi Arabia, Iran, the United States and Russia, opposing production caps. Once again, record numbers of lobbyists for the plastics industry attended. Taken as a group, the 220 fossil fuel and chemical industry representatives made up the largest delegation at the talks. However, a powerful group of 100 countries, including Canada, called for global limits on plastics production and voiced support for future negotiations.

The scene is being set for the failure of yet another false solution — this one potentially allowing the plastics tsunami to grow, destroying our lakes, rivers and oceans, overflowing landfills and destroying habitat for wildlife while the very industries most responsible for the climate crisis get not only to continue to pollute but to profit from the trade in plastics pollution.

Cryptocurrency blockchain trading to "save" nature

LYKKE called itself a cryptocurrency pioneer trading and designed a whole system of mangrove restoration funding powered by the blockchain and based on carbon credit incentives. The token, called Mangrove, because it represented one mangrove tree to be "saved," also entitled cryptocurrency investors to have "governance rights" in the mangrove restoration project. Before the exchange was liquidated after a hack in June 2024, its "digital tokens" were a new form of "natural capital" crypto trading.

The tokenization of environmental assets is the latest craze in the carbon and biodiversity trading world. As Frederic Hache of Green Finance Observatory reminds us, Bitcoin has had many serious issues in its short life, including extremely high price volatility, spectacular crashes, thefts and bankruptcies.[194] Now, the crypto universe has taken on a new "calling" — regenerative finance — aimed at promoting the UN Sustainable Development Goals, relying on the blockchain to address climate change and loss of biodiversity.

The regenerative finance movement aims to "fundamentally transform the governance of global common pool resources such as the atmosphere . . . by utilizing digital monitoring, tokenization of assets and decentralized governance approaches," according to the scientific journal *Frontiers*.[195] It is based on the concept of regenerative economics, which Wikipedia explains reflects the notion of the Earth as the "original capital asset." Regenerative finance is growing quickly and has the support of some environmentalists who are looking for alternatives to slow-moving governments to address the many environmental crises we face. It also has the support of the World Bank.

But as Hache points out, while the network Blockchain for Good Alliance boasts of many "positive impact" blockchain projects, the majority in its environment and climate category are linked with carbon and biodiversity offsets and trading. He says this is not surprising given crypto's inherent bias toward private sector initiatives and market-based solutions over government regulation.[196]

In 2020, World Rainforest Movement reported that the World Economic Forum launched the Earth BioGenome Project, intended to sequence and catalogue all plants, animals, fungi and single-cell organisms on Earth, as well as the Earth Bank of Codes, an online system that would use blockchain to store this data and register intellectual property assets as well as the rights associated with them. Aside from rendering government regulations obsolete, said the organization, blockchain systems render banks, notary offices, land registry offices, lawyers, security companies and any other trusted third parties similarly obsolete.[197]

There are many serious reasons to abandon blockchain as a potential solution to the climate crisis. Hache notes that the crypto energy footprint is already enormous: its CO_2 emissions are estimated to be on par with Greece and it consumes more electricity than Argentina.[198] It also generates massive electronic waste. Regenerative finance further entrenches the idea that conservation should be privatized and profitable and that offsets and nature trading are part of the solution.

"The libertarian political foundations of crypto assets mean that blockchain-based initiatives are inherently biased against government regulation and in favour of economic incentives, private-sector initiatives and market-based solutions," writes Hache. "When applied to environmental policies, this can translate into implicitly supporting a conceptual shift in conservation towards profitability criteria and market-based approaches."[199]

Indeed, researchers at the University of Pennsylvania underline that the origin of Bitcoin is libertarian, prizing "free" markets over government regulations. Cryptocurrency promoters demographically are millennials, men, politically conservative and "right-wing anti-establishment," and some hope to replace the banking industry with software. In a January 2025 report, they noted that President Trump claimed that cryptocurrencies could "define the future" and wanted them "mined, minted and made in the USA," promising to make America the "crypto capital of the planet."

On March 2, 2025, Trump announced on his social media platform that he would create a US Crypto Strategic Reserve, and a day later, the US Treasury Department announced it would no longer enforce its anti-money-laundering law. A month earlier, Argentina's President Javier Milei tweeted about the crypto coin $LIBRA, encouraging his followers to buy it on the promise it would help small business start-ups. Within hours, $LIBRA's price shot up, reported the BBC, but then nosedived, causing severe losses for most of the people who invested in it.[200]

Trump and his family have even launched their own crypto business called World Liberty Financial.[201] And at the end of March 2025, Donald Trump's two sons, Eric and Donald Jr., announced they were investing in American Bitcoin, a Bitcoin-mining company. "We are a hard-asset family. I'm a hard-asset guy," said Eric.[202] The family is making a fortune on these investments: *The Wall Street*

Journal reported that the Trump family amassed US$5 billion in paper wealth on September 1, 2025, the day after its flagship crypto venture opened trading of a new digital currency.[203] Ex–Goldman Sachs analyst Nic Puckrin warns that Elon Musk wants the US Treasury to move all of its transactions onto a blockchain that underpins cryptocurrencies like Bitcoin.

American journalist Chris Gay argues that, in spite of the crypto market being unregulated and extremely volatile, it is an increasingly significant force in financial markets and that systemically important financial institutions, and even pension funds, are increasing their crypto holdings. In a March 2025 editorial, Gay reported that the crypto market's total capitalization peaked at US$3.9 trillion at the end of 2024 and is projected to reach US$12 trillion by 2030. He warned about an unregulated "Wild West crypto market" meltdown.[204]

The world of cryptocurrencies is no place to entrust the protection of the Earth. Green Finance Observatory is calling for regulation of the environmental impacts of crypto assets and oversight of the environmental claims of its proponents. Supporting a system that would protect nature by further dividing it into "accountable" units that are then moved over to the libertarian, largely unregulated world of crypto to be collected as data so that they can then be financially assessed and traded for profit, is simply dangerous.

It's clear that the climate and biodiversity crises will not be solved by mimicking the growth-obsessed, profit-dominated strategies of the financial markets. The revolution we need will not come from the natural world being bought, sold or traded.

CHAPTER THREE

THE FIGHT FOR BLUE GOLD

There are the people who will buy the water when they need it. And the people who have the water to sell it. That's the blood, guts and feathers of the thing.

— T. BOONE PICKENS, LATE AMERICAN BUSINESS MAGNATE

On the day in 2019 trucks took away then 23-year-old Meghan Campbell's cows, she wept. Meghan had helped care for them at her family's 800-acre dairy farm in the Murray River district of Australia since she was a child and hoped to be the third generation to take over the farm. But her father ran out of options. Holding no water rights himself, he had to buy water on the spot market and drought had sent the average price of Murray–Darling water up 139% in just one year. "That's it, darling," he said to Meghan. "We're out." As the last truck pulled away, one of the cows Meghan had nurtured poked her head over the tailgate and looked back. Its name was Hope.[205]

Meghan's story is part of a *Bloomberg* exposé of how private investors are commandeering public water for profit in many parts of the world. The water crisis in Australia stands out for two reasons. One is that it is the world's driest inhabited continent, and it has experienced sequential droughts for decades. Many lay the blame on climate change. But a March 2025 study from four Australian universities found that declining streamflow in the Murray–Darling

Basin due to water diversions for agricultural farming and other commercial uses is just as big a factor.[206]

The other reason is that, in 1994, the Australian government, concerned about the growing strain on the Murray–Darling Basin, converted water licenses to water rights, setting up what would become the world's largest water market. The intent was to incentivize farmers to conserve water so they could make money selling what they didn't use. But soon, the smaller family farms were bought up by large-scale industrial farms, big banks and asset managers. Then the international water hunters moved in.

Australia has spent the past three decades building the world's most advanced water exchange, handing over significant control of its water to the market, reported *Bloomberg*. In 2025, it was valued at over US$10 billion. Every year, its water traders wheel and deal nearly 8,000 gigalitres of water — enough to supply the population of France for a year — at the value of over US$2.7 billion. "For deep-pocketed financial institutions and agribusinesses, many based overseas, water trading has been a bonanza. Big investors have wielded substantial financial clout to extract more water, and more profits, at a time when the asset is increasingly precious amid climate change and rising agricultural demand. They are abetted by financial firms that provide liquidity to the exchange, connect buyers and sellers, and earn substantial returns arbitraging the water market."

However, what might be good for investors has been a disaster for communities.

Indigenous people and farmers have seen their water disappear and the environment has been gutted. A 2021 700-page report by Australia's major antitrust regulator found that the water market is rife with opportunities for abuse. Today, a wine grower or cotton exporter can order water on a cell phone while water that is too expensive to buy flows past family farms closing all over the country. It turns out trading water has been tantamount to transferring wealth and is "sucking Australia dry," stated the report.[207]

The privatization, commodification and financialization of the planet's dwindling water supplies is a major and underreported story.

A planet running out

Our planet is running out of accessible clean water. We humans have seen water as a resource for our convenience and profit and have ruthlessly helped ourselves to its freshwater heritage to build our cities, run our factories and grow our industrially produced food. We have diverted water from its watersheds, paved over waterways, dammed our major rivers to death and greedily pumped groundwater far faster than it can be replenished by nature. We have polluted our lakes, rivers and aquifers, many of which are also being destroyed by saltwater intrusion due to our carelessness.

In his 2024 book, *Planet Aqua*, American writer Jeremy Rifkin asks us to imagine the incredible hubris of believing that our species could sequester, pacify, propertize, commodify and hold dominion over all the waters of the planet for our exclusive utilitarian uses. "Yet, that's exactly what we set out to do six thousand years ago with the rise of urban hydraulic civilizations around the world." Rifkin says we have got it wrong: we do not live on a land planet. In reality, we live on a water planet, and now the Earth's hydrosphere is taking us into a mass extinction event as it searches for a new normal.

"Our planetary hydrosphere, which animates all life on Earth, is rebelling in the wake of a global warming climate, unleashing blockbuster winter snows, biblical spring floods, devastating summer droughts, heatwaves and wildfires and deadly autumn hurricanes, wreaking havoc on ecosystems, infrastructure and society. While fossil fuels lit the fuse, it's the hydrosphere that's now ringing the death knell."[208]

An international group of scientists announced in 2017 that to keep pace with the global demand for food, crop production needs to grow by 50% by 2050, which will require an increase in water for irrigation by 146%.[209] Where is this water to come from? Another international research team studied every one of the planet's 3 million rivers over 35 years and published their shocking findings in December 2024: Rivers everywhere are in crisis. Nearly half of the largest downstream rivers saw a yearly drop in water flows — some, such as the Congo in Africa, the Yangtze in China and South America's Plata, are experiencing dramatic declines. A small portion of smaller rivers in mountain areas saw stream flow increase, but this is likely due to melting glaciers and snowpacks.[210] NASA satellites track the rivers of the world and have noted a major drop in global freshwater since 2014.[211]

Lakes are in no better shape. Scientists from the US, France and Saudi Arabia examined the Earth's 1,972 largest lakes and reservoirs and found that over half are shrinking. Some, like the Aral Sea in Uzbekistan, Lake Mead in California, the Dead Sea between Israel and Jordan, Lake Chad in west-central Africa, Lake Urmia in Iran and the Caspian Sea in Eastern Europe and West Asia, are disappearing. Calling the results "staggering," report co-author Balaji Rajagopalan with the University of Colorado Boulder said that lakes are in trouble globally, and that has implications far and wide. "If a good chunk of freshwater lakes are drying, then you're going to see the impact come to you one way or the other, if not now, in the not-too-distant future." There were some lakes experiencing water storage increases, but they tended to be found near melting glaciers and permafrost.[212]

Yet another disturbing 2024 study reported that groundwater levels around the world are dwindling quickly, often at accelerating rates. Roughly half the world's people get their drinking water from groundwater, said the team from the universities of California, Santa Barbara, and ETH Zurich, with India, Pakistan and the United States in significant jeopardy. (Lack of information for China, Russia and large parts of Africa leaves gaps in the study.) In California, growers treated underground aquifers as unlimited sources of water until the state finally started to regulate their access. By then, unfettered pumping had already left more than 20 of the state's groundwater basins — lifelines for many community drinking water systems — "critically over-drafted."[213]

This ecological crisis has huge human implications. The UN reports that by 2030, demand for water will outstrip supply by 40%, and that over 70% of people now live in countries that are water insecure or critically so.[214] The World Health Organization says that one in three people globally do not have access to safe drinking water and half do not have access to safely managed sanitation. When the coronavirus pandemic hit, more than 4 billion people could not follow the simple universal instruction to wash their hands with soap and warm water.[215] In his 2023 book, *The Last Drop*, British environmental journalist Tim Smedley warns that cities and even countries are "flirting" with Day Zero — the day its water system goes dry. In a review, Canada's *The Tyee* says the book describes, in horrifying encyclopedic detail, the effect of water stress in countries all over the world.

"Cape Town in South Africa faced a Day Zero a few years ago, as did Chennai in India and São Paulo in Brazil. According to the World Resources Institute, 33 countries will face extremely high water stress by 2040. High-stress countries like Mexico, Spain and Turkey consume 40 to 80% of their available water each year; countries like South Africa, Saudi Arabia and India consume over 80% of their water, making them extremely highly stressed."[216]

Clearly, the global water crisis is threatening not just humans, but the Earth itself, causing great suffering and accelerating the loss of biodiversity. In January 2025, the International Union for Conservation of Nature published the largest global assessment of freshwater animals ever conducted. Fully one-quarter of freshwater animals are at risk of extinction due to water pollution, overfishing, excess nutrient loading, land conversion for agricultural use, invasive species, water extraction and major dams that block fish migration routes.[217]

Water becomes a "good"

For the last half-century, there has been a mighty contest over the planet's dwindling freshwater supplies between those who say water is a commodity to be put on the open market like oil and gas and those who say it is a human right and public trust that must be governed by democratic principles. Large private water operators and investors knew before most of the public that, in a world running out of clean water, those who control it will be both wealthy and powerful. More recently, the same forces that have ravaged forests, bottom-trawled the oceans, polluted the atmosphere and carved out mountains for gold are seriously looking to the financialization of water as a new frontier for profit.

The commodification of water has happened in stages. Water "in all its forms, including ice and snow" was included as a "tradable good" in the General Agreement on Tariffs and Trade, the 1989 Canada–United States Free Trade Agreement and, several years later, the North American Free Trade Agreement and is included in many other global free trade and investment agreements.

At the 1992 UN freshwater conference in Dublin, for the first time, water was referred to as a "commodity" and an "economic good." The next year, the World Bank called for water to be considered an economic commodity with an emphasis on efficiency, financial discipline and full cost recovery, even from poor countries. Private water utilities such as Suez and Veolia were beginning to move into cities of the Global South and wanted guarantees that they would make a profit.

Public water services get targeted

The privatization of water services started in France in the early 1980s, when neoliberalism was on the rise and political leaders in Europe started privatizing their transportation, energy and water services in the belief that the private sector could deliver better services at lower costs. Paris privatized its water services in 1987, as did many other French cities in that decade. In 2009, after years of poor service and damage done to unprotected water sources, Paris took its water services back under public management and created a public water utility called Eau de Paris. Soon other French cities followed suit. According to a French public agency, between 2008 and 2023, the proportion of the population of France served by a public operator rose from 38% to 48% for drinking water and 56% to around 66% for sanitation.[218]

In 1989, Great Britain's Prime Minister Margaret Thatcher sold the entire package of public water companies in England and Wales to private water corporations, including Thames Water. The damage this has caused has been well-documented, with calls coming from across the political spectrum to remunicipalize their water services. UK water bills have soared over 360% since privatization — double the rate of other goods and services — and decades of poor regulatory oversight have allowed company owners to pay themselves billions of pounds in dividends instead of using the profit to maintain infrastructure, according to *Bloomberg*. "UK water utilities have enjoyed a rock-steady stream of income from bill-paying households and businesses, but many of them are laboring under heavy debts racked up by the international funds that own them."[219] The largest, Thames Water, was on the brink of running out of money when it was rescued with an almost US$4 billion government bailout in February 2025.

The private water companies have used England's rivers as open sewers. In 2023 alone, they increased their sewage dumping by 105%. Thames Water dumped at least 72 billion litres of raw sewage into the Thames River in just two years, 2020 to 2022, reported the BBC, and another 300,000 hours' worth of effluent discharge in 2024, up another 50% from the year before.[220] In his first interview since taking office in July 2024, British environment secretary Steve Reed said, "Every single river in England today is polluted. The public are quite rightly furious that they have to worry about letting their kids splash about in the river, for fear of what they might catch because it is polluted."[221]

As often happens with water privatization, the British public has lost political control of these water companies. England's nine water and sewage companies are more than 90% owned by overseas investors, with many hailing from tax havens such as Bermuda or nations known for human rights abuses, reported British *The i Paper* in 2023. As well as pension funds, foreign corporate investors, including BlackRock, come from Australia, Qatar, the US and China. Three are wholly or partially owned by China, including the nation's largest supplier, Thames Water. As a result, more than half of England's water bills help fund Chinese investment firms. The poor performance of these private water companies can be attributed to the fact they are operated in the interests of transnational corporations and foreign sovereign wealth funds rather than the British public.[222]

The 1990s saw a wave of water privatizations across Europe and the United States, and the World Bank started to promote private water services in the Global South. Almost universally, water rates went up, the quality of service went down and source waters suffered from neglect. The United Nations special rapporteur on extreme poverty and human rights found that "privatization often involves the systematic elimination of human rights protections and further marginalization of the interests of low-income earners and those living in poverty."[223]

The backlash from local peasant and Indigenous communities was fierce, and projects in many cities of the South, from Cochabamba and Buenos Aires to Dar es Salaam and Lagos, were held off, cancelled or defeated by strong civil society action. After an intense civil society campaign and referendum, Uruguay became the first country in the world to amend its constitution to ensure the human right to water. Cities in the Global North followed suit: Hamilton, Atlanta and Berlin are among more than

300 towns and cities that tried privatization but brought their water services back under public management — a huge victory for the global water justice movement.

However, the major water companies have not given up. According to Public Services International, a global union representing 30 million public sector workers in 154 countries, the civil society pushback against water privatization forced the world's two biggest private companies, Suez and Veolia (which still run the private water systems of 284 million people), to merge and redirect.[224]

In 2022, Veolia essentially completed a "hostile takeover" of the major assets of its competitor, strengthening its own position in the water service market, giving it a de facto monopoly on certain technologies the company believes will become profitable as the climate crisis heats up. Veolia now has around 230,000 employees with an annual turnover of close to US$45 billion. Suez is substantially downsized but intends to specialize in contracts linked to development in Africa, supported by international financial institutions.

The union says that Veolia is repositioning itself as a provider of "climate solutions" that will allow it to capture part of the climate funding from governments and financial institutions and is an illustration of "crony capitalism" at the highest level of the French state. Even though these companies are private and have been rejected by many French cities, including Paris, the deal was supported by the French government and made possible by the country's major public financial institution, Caisse des Dépôts, often referred to as the investment arm of the French state.

American watchdog Corporate Accountability says this merger has set the stage for the monopolization of the water sector, whereby an even smaller group of wealthy individuals and institutional shareholders of the Global North will expand its reach into the Global South, worsening an already unjust status quo.[225]

The "new Suez" was purchased by a consortium of investors including the Caisse des Dépôts; Meridiam, a Paris-based global asset manager specializing in public-private partnerships that works with the European Investment Bank and has an asset portfolio of US$20 billion and was Veolia's partner in the takeover of Suez; and Global Infrastructure Partners, an American investment fund with deep ties to the World Bank and assets of US$100 billion making equity and debt investments in several sectors, including water. In early 2024, Global Infrastructure Partners was bought by the assets giant BlackRock, bringing its formidable global footprint deeper into the private water sector.

A partnership between Meridiam and a fund set up over 15 years ago by governments to fight the climate crisis gives even more power to private asset investors who promote water privatization in the Global South. The Green Climate Fund (GCF) was established in 2010 at the climate COP in Cancun for wealthy countries to assist developing countries in dealing with the fallout from climate change. With a portfolio of US$13.5 billion, it is considered the world's largest climate fund and was meant to be a kind of reparation payment from countries of the Global North in recognition of their overwhelming responsibility for climate damage in poor countries.

But the fund has been criticized for its growing partnerships with private corporations and asset managers. "Climate change offers businesses an unprecedented chance to capitalize on new growth and investment opportunities that can protect the planet as well. GCF employs part of its fund to help mobilize financial flows from the private sector to compelling and profitable climate-smart investment opportunities," it says.[226] As of 2024, the fund has committed US$15 billion to 270 climate projects worldwide, of which US$5.5 billion is earmarked for 63 private-sector projects.

In October 2024, the Green Climate Fund and Suez partner Meridiam convened an investor forum in Paris where they promised to unlock billions of private sector capital for investors "seeking growth opportunities" in the water business the Green Climate Fund says that facing increasing demand for water, the global water business offers great investment opportunities and calls for a "paradigm shift" toward greater investment into a new "water asset class."

It is the poor who will pay for this water grab. Too often, the political and business elite in poor countries of the Global South use the desperate need for money to provide water services as a green light to make deals with these water barons. Corporate Accountability says mergers like this are already rife with issues of transparency and accountability. But when the "product" is a human right, the consequences of such a deal can be orders of magnitude worse.[227]

Water markets commodify lakes and rivers

Water rights, water trading and water markets are another form of water commodification. Water rights give a user guaranteed access to a local water source. They are either inherited or granted when a government converts a water license to a right, giving the new "owner" more discretion in the use of the water, including separating it from the land and selling it. The countries with the most advanced water markets are Australia, the US and Chile.

Chilean dictator Augusto Pinochet not only privatized all the water services in his country, he created a Water Code giving wealthy landowners the water rights on their property, which they were then free to put on the open market. This led to the auctioning off of entire watersheds to foreign mining and agribusiness corporations, who were then free to dump their pollution in "their" water. As Sara Larraín of the environmental organization Chile Sustenable reported, once the water rights are sold to private investors, reallocation of water happens by means of a water market. The ownership of Chile's water has become concentrated in the hands of the hydroelectric, mining and agribusiness export sectors, with devastating impacts on the environment.[228]

Chile, historically a water-rich country, is suffering from the longest-lasting drought it has experienced in almost 1,000 years, according to studies at the University of Chile. Climate change, water abuse and theft by agricultural companies and overexploitation by mining and forestry companies have diverted water from poor and Indigenous communities, and hundreds of thousands rely on water transport trucks for drinking water. A devastating April 2024 report by the Chilean Center for Climate and Resilience Research suggests the country is nearing Day Zero within decades. The water crises in the central and southern basins are approaching "permanently high and extreme levels of water stress," and demand for water already far exceeds availability.[229]

Yet the destruction of Chile's water is escalating as Chilean and foreign investors flock to extract that country's vast reserves of lithium, used to make batteries for electric cars. The majority of the world's lithium is produced on salt flats, lithium-rich brine deposits found in subterranean reservoirs, predominantly located in the "Lithium Triangle" — a region spanning the borders of Bolivia, Argentina and Chile, containing about 56% of known reserves. Lithium-rich saltwater is extracted from underground reservoirs and held in large evaporation ponds that, over time, increase concentration of lithium in the brine. The process is water-intensive, as fresh water is used to filter the salt brine at the surface. The production of one ton of lithium requires about 2.2 million litres of water, reported Greenly, a company that advises other companies on their carbon footprint.[230]

SQM, a Chilean chemical company, has been operating a lithium mine in Chile's Atacama Desert for decades. However, in 2015, as the global demand for lithium started to spike, North Carolina's Albemarle, the world's largest lithium producer, obtained a concession. Annual production at its mine has increased from 22,000 tons to 84,000 since 2016, and as of 2025 the company employs more than 1,000 in Chile, up from 250 in 2020. The company anticipates explosive expansion. The operations from this American company alone consume around 21 million litres of water every day, using 65% of the region's water, resulting in severe water shortages for the local, mostly Indigenous, community. And now Chinese giant electronic carmaker BYD is bidding for a concession in this desert as well.[231]

The Bolivian government is racing to attract foreign investment in its massive untapped lithium reserves, particularly in the Uyuni salt flat in the country's Andean southwest, believed to contain the largest lithium deposit in the world. In late 2024, it signed a US$1 billion deal with China's CBC Investments, a subsidiary of Chinese company CATL (the largest EV and energy battery manufacturer in the world), to build two plants on the flats. A year earlier, Bolivia signed a US$450 million contract with Russia's Uranium One to mine in the same area. Other foreign corporations and investors from France, Italy and Australia are lining up to mine Bolivia's "white gold," seeking proprietary rights over the already endangered water sources of this biodiversity reserve.

A March 2025 study published in the journal *Communications Earth & Environment* reported the amount of freshwater locally available for lithium extraction in the Lithium Triangle is about 10 times lower than previous estimates. Global lithium demand is expected to grow by 40 times by 2040 and could outpace the annual rainfall in the region.[232]

In the western United States in the late 19th and early 20th centuries, water promises were used to encourage settlement and industry by allocating permanent water rights that could be passed on to future generations. In recent decades, water rights holders in several US states have claimed their right to decouple the water from their land and sell it on the open market.

One of the many examples is the small town of Cibola, Arizona, which, in 2013, sold 500 acres of agricultural land to Greenstone, a private water company backed by global investors, only to find that several years later, the company had sold the water rights that were tied to the land to Queen Creek, a fast-growing suburb of Phoenix, for a US$14 million profit. More than 2,000 acre-feet of water from the Colorado River, once used to irrigate farmland, is now flowing through a canal system to homes almost 300 kilometres away. In 2021, Queen Creek paid US$30 million for 5,000 acre-feet of water from another basin west of Phoenix and is scouting for more water sources to buy.

In its series on water grabs, *Bloomberg* said the explosive growth in water-scarce areas of the American West has set up "water battles" that will see prices rise dramatically as more farmers sell, leaving local communities high and dry. A lawyer for Greenstone said that the problem is that society hasn't allowed markets to operate with regard to water "like we do to other resources." Holly Irwin, a local town supervisor, told *Bloomberg* she was worried that everyone with water rights will transfer them, "and there will be nothing left." She added they don't know the limit to the price of water yet. Her colleague Paul Gardner agreed — "We are about to find out" — and predicted the costs will be astronomical.[233]

This water company is not alone, reported non-profit news outlet *Truthout* in 2024. Water Asset Management, a New York–based hedge fund, owns more than 6,200 acres in Arizona's Maricopa and Mohave counties and is a major landholder in Colorado's Grand Valley. The company's president, Matthew Diserio, called water an undervalued asset and "the biggest emerging market on Earth . . . a trillion-dollar market opportunity." *Truthout* reports that water corporations are plundering the water of the US west amid the worst drought in 1,200 years. "Water management companies" are buying up farmland throughout the Southwest, drilling to new depths to find the water on their purchased land with the intention of selling the water rights at a profit to thirsty cities and suburbs elsewhere.[234]

The situation is even more dire in California, where "financier-farmers" — investors from New York, Toronto and Zurich — have bought up large parcels of land and dug deep into the earth to extract the water. Since 2010, according to *Bloomberg*, six major asset companies have quadrupled their farmland, dramatically depleting water tables and driving the price of water up. "Wall Street has come to

Woodville," says the report, referring to a small rural town in the northern part of the state. Banks, pension funds and insurers have been turning the state's scarce water into enormous profits, leaving people with less to drink, or no clean water at all. And it is happening all over the world. Globally, large investors and agribusiness have snapped up about 163 million acres of farmland in more than 100 countries — more than 54 million acres in Africa alone — in the past 20 years.

New York–based African Agriculture Inc. is an investment company growing 740 acres of water-intensive alfalfa inside a desert nature preserve in Senegal. The company plans to expand its operations to 20,000 hectares (almost 50,000 acres) to produce a crop destined for export to Saudi Arabia to feed its livestock. *Bloomberg* reported that the alfalfa farm draws its water from Lake Guiers, the only freshwater reserve in the country that also supplies half the water for Dakar, its capital. According to the World Bank, Senegal is moving dangerously close to having insufficient water to meet its basic needs. But like many countries being invaded by these water hunters, it is "resource-rich, cash-poor, weak in governance and vulnerable to exploitation."

The area is home to 35 villages of nomadic people who now have to walk for days at a time to get water for their livestock, as the farm administrative buildings are encircled with barbed wire and the farmland is surrounded by a trench to keep out the livestock. A chief of one village told *Bloomberg* the reserve and their way of life will disappear if the second farm is granted. "And for what? So a foreign company can sell alfalfa to feed cows and racehorses in Saudi Arabia?" The article also quotes Ousmane Aly Pame, a university professor who also runs an anti-desertification non-profit on the Senegal River: "Extracting resources and leaving Africans with no food and no future is exactly what happened in colonial times."[235]

African Agriculture openly touts cheap water from Africa as an incentive for investing in its company. It is seeking capital to open another alfalfa operation of up to 400,000 hectares in Mauritania, located in the Sahara Desert and itself in water crisis, to help it fulfill an offset agreement to plant a million trees to generate carbon credits in Niger, using groundwater for irrigation.

"The land grab has given rise to a grab of an even scarcer global commodity: water. In a bid to ensure thriving investment portfolios, some of the world's largest financial entities have amassed control over lakes, rivers and underground aquifers in places from California to Africa, Australia to South America, giving them outsized roles in managing an endangered resource that's the basis of life on Earth. The trend has contributed to shifting hydrological patterns that stand to permanently disrupt communities' access to freshwater. Local populations are paying the price in drained wells, high water bills and contaminated water supplies."[236]

Many water transactions that commodify public water are between large private corporations and public utilities whose mandate is the provision of water to local homes and businesses. Elon Musk's rocket and spacecraft company, SpaceX, purchases large amounts of water from the public utility board in Brownsville, Texas, to suppress fire, dust and debris when it launches from its nearby test site. The company claims it uses 180,000 gallons (almost 700,000 litres) with each launch of its Mars space rockets. Local environmental group Save RGV estimates over 18 launches, SpaceX used at least 3.24 million gallons (over 12 million litres) of municipal water, the vast majority of which ends in a cloud of steam. The utility board, which has raised water rates to the public to deal with the prolonged drought in the area, refuses to disclose the purchase agreement with the company.[237]

Artificial Intelligence (AI) uses massive amounts of water to cool its densely packed servers and computing hardware racks. In 2022 alone, Microsoft's water consumption reached nearly 1.7 billion gallons and Google used 5.56 billion gallons — 800 times as big as the Lincoln Memorial Reflecting Pool in Washington, reported Veolia.[238] A 2025 investigation by non-profit SourceMaterial predicts that Amazon, Microsoft and Google plan to increase the number of data centres they operate, as cloud computing and AI cause a surge in the world's demand for storage. As of early 2025, there were 342 data centres, some as big as multiple football fields, operating in water-starved Texas alone, most accessing public water sources.[239]

In 2025, *Bloomberg* reported that American AI centres are drawing water from areas that need it the most: two-thirds of new AI data centres are in parts of the US already in the grip of high water stress. "Each time you ask an AI chatbot to summarize a lengthy legal document or conjure up a carbon squirrel wearing glasses, it sends a request to a data center and strains an increasingly scarce resource: water."[240] Google's data centre in The Dalles, Oregon, is called a "water vampire" by residents as it consumes almost one-third of the town's water supply. In 2022, the company did a water trade with the public water utility, ceding some of its water rights from industrial land it owns in order to have increased access to public water. Google is planning two more data centres in this small town of 16,000. Asked about the harmful environmental effects of AI in Washington, DC, reported the Institute for Policy Studies, former Google CEO Eric Schmidt said we might as well continue to invest in AI because "we are never going to meet our climate goals anyway."[241]

AI is just the latest form of public water capture by private interests who are adding in the creation of water markets. Food & Water Watch says there are three threats posed by water markets: water hoarding, where investors can buy up water claims to speculate on the price of water at the expense of nearby residents and the environment; price gouging, where private water markets significantly increase water bills for consumers while encouraging unsustainable water-use practices and deteriorating water quality; and the misuse of water supplies, where water is allocated to the highest bidder, not to the highest public value for people and the planet.

Water markets and water trading also open the door to foreign investors, giving transnational corporations and asset managers political clout in other countries. Foreign ownership in Australia's water market has steadily grown since it established its water market three decades ago. According to the Australian government, foreign investors — mostly in the mining and agriculture sectors — hold about 12% of the water in the Murray–Darling Basin. The largest is a Canadian pension fund that manages the pensions of several police and defence forces and has poured billions into Australia's food, wine and cotton industry. Other investors hail from China, the US and the UK.

Maryanne Slattery, former director of Australia's Murray–Darling Authority, told *INTHEBLACK* that water markets only serve those with deepest pockets: "Water does not move to its highest value use for the community, the economy or the country. It moves to whoever is prepared to pay the most." Patrick Viljoen, sustainability adviser with CPA Australia, noted that moving water around is like robbing Peter to pay Paul. Water trading lends itself to price gouging, hoarding or, at the most extreme, water wars, he writes on his organization's website.[242]

Yet the big conservation NGOs, financial sectors and the World Bank all tout water markets and water trading as a key solution to the water crisis. The Nature Conservancy openly promotes Sharing Investment Partnerships that use investor capital to acquire water-use rights, which can then be "reallocated to depleted freshwater systems, or sold or leased to other water users to generate financial returns for investors."[243]

There is another serious issue here: rapidly escalating water transfers. Raw waters being moved from lakes, reservoirs and rivers via canals, pipelines and tunnels are carrying invasive species, and the threat has passed largely under the public radar. In a study published in March 2025, researchers at Britain's Newcastle University School of Natural and Environmental Science reported that invasive species pose a major threat to local biodiversity and can overuse resources and destroy habitats. They cited the "dirty dozen," including zebra mussels native to Russia, as particularly harmful to Great Britain. Giant African land snails that consume at least 500 different types of plants have invaded the US, from Florida to New York City. Raw water transfers lack proper regulations and surveillance, and are acting as a conduit for invasive species. The researchers warned, "We need to take the risk seriously."[244]

Bottled water privatizes local water sources

Another form of water privatization and commodification is bottled water. The issue of bottled water has already been raised in the section dealing with its plastic footprint, but the industry has a dangerous water-taking footprint as well. Huge water corporations such as Nestlé (which owns 47 brands of bottled water, including Perrier, Pure Life and Poland Spring), Danone, Coca-Cola, PepsiCo and China's Nongfu Spring drain local springs and aquifers, paying a pittance to the local communities whose water sources they destroy. When they get access to these springs and groundwater sources by government license or water markets, they essentially privatize and control the local water source, impacting the larger watershed.

Swiss food and bottled-water giant Nestlé had a 2024 profit of over US$12 billion, slightly down from over US$14 billion the year before.[245] While profit figures for its water division are not listed separately, in May 2025, it was reported that the company sold it to the Volvic Group for US$12 billion.[246] The company's water division has operated in every continent except Antarctica, and controversy and opposition to its presence have followed it everywhere. Nestlé bought access to the mineral waters of São Lourenço, a water park in an area of Brazil famous for its medicinal properties, where people have come for many years to heal. In 2001, charges were brought against the company as its taking of 53,000 litres of water an hour from a well 150 metres deep had demineralized the water, polluted nearby streams and driven away tourists.

In 2013, thousands of Pakistanis demanded that Nestlé stop draining groundwater from poor, water-parched villages to bottle its Pure Life water. In 2016, two years after Nestlé set up operation in Manderegi, Nigeria, residents struggled with chronic water shortages as the stream that served as a source of water supply was contaminated with wastewater pumped from the Nestlé factory, reported the International Centre for Investigative Reporting.[247]

Nestlé has made billions selling the waters of North America and left a trail of broken job promises, environmental degradation and dwindling water supplies. In 2003, a Michigan court ruled that Nestlé was solely responsible for draining the Dead River watershed from which it was pulling 1,500 litres every minute. Over widespread opposition from the local community in Fryeburg, Maine, the company won the right in court in 2016 to take up to 830 million litres of water a year from a local spring for 45 years. In 2018, Nestlé siphoned over 170 million litres of water from Strawberry Creek in the San Bernardino Forest, two hours east of Los Angeles, California, paying nothing for the water and leaving creek beds in the area bone dry.

In the Guelph area of Ontario, Nestlé has been extracting 4.7 million litres of water every day for decades from springs on the Grand River watershed, situated near Six Nations, an Indigenous community where thousands still have no running water in their homes. Facing fierce opposition, Nestlé sold its North American operations in 2021 for US$4.3 billion to One Rock Capital — a large private equity firm. The Guelph-area bottled-water operations have continued under the name BlueTriton. In the two years after acquiring Nestlé's water operations, reported *Bloomberg*, BlueTriton increased water pumping in at least six states: Florida, Maine, Colorado, Michigan, Texas and South Carolina.

Arlene Slocombe with the Ontario environmental organization Water Watchers criticized both seller and buyer: "Nestlé's motivation is clear: to shed itself of its responsibility for the plastic pollution and environmental degradation its water extraction and bottling has caused and the damage these scandals have done to their brand and bottom line. It is also clear that a private equity firm, freed of Nestlé's reputational responsibilities, will seek to cut expenses at the cost of the limited promises its predecessor made regarding environmental sustainability and community benefit."[248] Water Watchers and many others rejoiced when the company announced it was ceasing all bottled-water operations in Ontario in 2025.

Unfortunately, in mid-summer 2025, it was reported that White Wolf, a property management company that owns several bottled-water operations in Nova Scotia, had purchased the two vacated wells located in Aberfoyle and Hillsburgh without a permit from the Ontario government. It soon came to light that, in keeping with his desire to hasten resource extraction in his province, Premier Doug Ford announced he will no longer require that a new water-taking company seek a permit upon the transfer of property. Said Arlene Slocombe, "This proposal is reckless. Water permits should never be handed off like real estate deeds. Communities and ecosystems deserve a say every time water is taken from the ground — not just the first time. Particularly Indigenous communities from whose treaty lands this water is extracted while many still experience water insecurity."[249]

China's richest man, Zhong Shanshan, amassed his US$63 billion fortune drawing water from some of China's most ecologically sensitive lakes, mountains, springs and waterfalls and selling it to the country's burgeoning middle class, reported *Bloomberg*. His company, Nongfu Spring, has 12 collection sites in remote areas of China. In 2019, it extracted 33 million cubic metres (33 billion litres, enough to fill 13,000 Olympic-sized swimming pools) and is seeking more. Nongfu Spring's facility on Wanlv Lake in Guangdong province is the size of 50 football fields and is set to expand. Water levels in the area have receded so much that ancient relics from the Ming and Qing dynasties were uncovered from the lake bed. At the company's site in Mount Wuyi, a remote UNESCO World Heritage Site in Southeast China, tea farmers struggle to keep their crops alive, bamboo forests have dried up and tourists who arrived to see the region's majestic beauty are greeted by trickling streams, according to *Bloomberg*.

China is the world's largest consumer of bottled water, and sales are predicted to skyrocket in the next decade. In most Chinese cities, the tap water is not safe to drink. Nongfu sells water that comes straight from the source, which the company claimed in a nationwide campaign has greater health benefits than water purified in a factory. This has made Nongfu a wildly popular investment. For its next water source, Zhong has set his sights on the high mountains of Tibet, one of the wonders of the world, *Bloomberg* further reported, "Nongfu is working to bottle melted glacier water from the Nyang River, a sparkling blue-green waterway that winds through snow-capped mountains where endangered Bengal tigers and red pandas live." The Chinese government supports this travesty. The local Tibetan Communist Party chief plans to make bottled water a major industry. A billboard by the Wanlv Lake site declares the government's intention to develop a "water-based economy."[250]

Despite several recent catastrophic floods, Spain is suffering from severe water scarcity. Many farmers and villagers of Catalonia in northeastern Spain are angry that foreign bottled-water companies are taking the water from the springs and streams that used to be a source of their drinking water, forcing them to buy the water back in plastic bottles at supermarkets. *The Guardian* reported that in February 2024, the government declared a drought emergency, whereby 80% of the population of Catalonia, including Barcelona, faced fines if they used more than 200 litres a day for drinking, washing, cleaning and cooking. Water restrictions were also imposed on agriculture and industries but not on the bottled-water companies. Foreign water giants, including Danone, Nestlé, Coca-Cola and PepsiCo, operate in water-parched Spain. There are more than 160 water extraction licenses given to these companies in Spain, 27 in Catalonia alone.

Members of a local environmental group sat outside just four of the Catalonia plants, including Nestlé's, from 6 a.m. to 10 p.m. for nine weeks and counted the water trucks leaving the sites. They counted a total of 185 trucks each day. Multiplying the number of crates of water each carried, they calculated these four sites alone manufacture the equivalent of 1.8 billion litres a year. The private water companies are able to extract underground stores of crystalline high-quality drinking water, which are better protected from pollution than rivers and reservoirs, noted *The Guardian*, and a disaster-stricken population has found itself reliant on bottled water, sourced locally but owned by private corporations. Farmer Roser Albó Garriga told *The Guardian*, "We used to have springs bubbling up everywhere. I never dreamed I'd have to buy water to drink. Last year, we had to ask neighbours lower down the valley to give us tubs of water just so we could wash. So to see trucks drive away from here, full of bottles of water, breaks my heart."[251]

Mexico is facing its worst water crisis in over 30 years. Reservoirs are tapped out and much of the country's surface water is totally contaminated. Many cities, including Mexico City, are on the brink of Day Zero. Anger has grown in cities like Monterrey, where taps run dry but bottled-water plants are extracting billions of litres of water from public reservoirs and private wells. "It's not drought, it's plunder," is the rallying cry in impoverished urban communities and rural villages. Coca-Cola is everywhere in Mexico: on the front of mom-and-pop stores, on the sides of delivery vans and 18-wheeler trucks. It even appears stamped on the sides of bottles being drunk by toddlers, reported *Al Jazeera* in a February 2025 story. And it has huge political influence. Before becoming Mexican president in 2000, Vicente Fox was CEO of Coca-Cola Mexico.

At its site in San Cristóbal de las Casas, the company extracts over 1.14 million litres daily, leading to severe water stress where one in three people in rural areas have no access to running water and the 200,000 city dwellers receive running tap water only three hours a day. No one trusts tap water anyway, as it is contaminated with sewage pollution, so locals are forced to buy their own groundwater in plastic bottles. This corporate assault started in 1994, with the signing of the North American Free Trade Agreement that allowed private investors from the United States access to Mexico's water for the first time. Coca-Cola pays a pittance — US$155 — for each license to open an extraction site. Its water and soda are sold in the 20,000 convenience stores it operates in Mexico. One environmental activist likened the company to drug cartels that create large networks of small-time dealers.[252]

Water offsets are launched

As with carbon, biodiversity and plastics, there are now water offsets, a fairly recent addition to the way in which water can be traded.

Water offsets can and do have a positive role to play in some circumstances where new demands on local water sources are met with water savings by other users in the area. A new housing development, for instance, knowing it will put more pressure on local water, encourages or compensates other local users, such as industry, to reduce their water footprint in order to maintain the previous water-taking levels in the area. The promise is that the creation of new communities and businesses will have zero impact on existing water sources.

However, a number of companies are now perverting this original intent and are promoting water offset credits and trading much along the model of carbon offsets. The companies, instead of reducing their own water footprint, get credits for doing something for water elsewhere — usually in an impoverished country or community. Projects include restoration of wetlands, aquatic ecosystems, degraded watersheds and riparian forests. Britain- and Cape Town–based MarillionEx recently added water to its nature-trading portfolio. "The path to sustainability is a long one, and we're committed to walking it, one offset at a time," it proudly states on its website.[253] One of its affiliates, Green Earth Africa, noted earlier for its eucalyptus plantations, is also in the water-credits business: "Join the movement towards water sustainability by purchasing water credits to offset your water footprint and make a positive impact on global water resources," it invites.[254]

Circular Ecology, another UK-based carbon offset company, is also getting into what it calls the "water carbon offset business" and says that by obtaining their credits from drilling boreholes in Uganda, investors get to help improve the lives of women and children and advance the UN Sustainable Development Goals on water. BioCarbon, a Bogota-based for-profit carbon trading company, launched a program of "verified water credits" in April 2025. The company says that one verified credit represents one cubic metre of water preserved or sustainably managed per year, adding that water is a "critical cross-cutting ecosystem service that, until now, has not been economically valued."[255] While it is possible that the intent behind these projects is genuine, and some may even come to pass, the reality is that the model allows a company to continue to abuse or overextract water in its own operations.

In 2007, Coca-Cola partnered with the Nature Conservancy to launch a water offset program, promising to replace "every drop" of water used in its beverages. Its offsets included providing basic services in Africa through well digging, water metering and watershed restoration through tree planting and high-tech irrigation projects. Coca-Cola promised to "bring safe drinking water and sanitation to people in the communities we serve" and designed a credit measurement system called the Water Replenishment Report.

However, a 2018 investigation by a team of independent American journalists found that the nearly 2 billion litres of water the company claimed to offset in 2015 covered little more than a small percentage of the water in each bottle and did not take into account the large amount of water it takes to produce each bottle in the supply chain. In other words, 99% of Coca-Cola's water use was not offset at all, which did not stop the company from declaring that it had reached its "water neutrality goal."[256] Another broken Coke promise was made public at the end of 2024 by the NGO Break Free from Plastic. Coca-Cola quietly abandoned its

2022 pledge to achieve 25% reusable packaging by 2030, a move the environmental group called "a masterclass in greenwashing."[257]

Water pollution trading

Another form of water markets is water pollution trading, also called "water quality trading" or "water nutrient trading" by proponents. Water pollution trading is a market-based approach to protecting lakes and rivers overloaded with industrial, human and animal waste. It allows a polluter — for instance, a factory farm — to maintain pollution levels that are higher than allowed by local regulations, by selling credits to another source. That source, in turn, uses the credits to reach its own regulatory compliance. Many of the projects centre on lakes that are contaminated with eutrophication, nutrient overload that depletes oxygen in the water.

One of the world's largest and oldest projects is in Chesapeake Bay, a major drainage basin covering six American states. The US Environmental Protection Agency (EPA) has warned that nutrient overload on the basin is drastically increasing. Claiming to help offset this ecological damage, the agency supports water quality trading "as a market-based approach," providing economic incentives for pollution reductions. "Trading can allow one source to meet its regulatory obligations by using pollutant reductions created by another source with lower pollution controls."[258]

World Resources Institute, another influential global environmental organization with close to 2,000 staff worldwide, supports water pollution trading and is very clear about its purpose. Some pollution sources are regulated but others are not, it says, leading to "cost differentials," thus creating an "ideal environment" for trading. "Trading allows sources with higher pollution control costs to purchase pollution reductions from sources with lower costs. Those with higher costs can save money, while those with lower costs can earn new revenues."[259]

In other words, water pollution trading can allow a company that doesn't invest in meeting its local environmental standards to take advantage of companies that have lower local standards and get tradable credits from the deal.

Washington-based Food & Water Watch released a scathing 2015 report showing that water pollution trading has given industry a way around EPA water protection rules. In Pennsylvania in 2014 alone, agriculture-related operations — mostly factory farms — bought credits to dump more than 800,000 kilograms of nitrogen and over 50,000 kilograms of phosphorus into local waterways. The system actually sees factory farm manure being trucked from one impaired watershed to another to generate profits, the authors reported.[260]

Canada's Lake Winnipeg, the 10th largest lake in the world, is in grave danger due to the excrement load poured into it every day from the hog, cattle and poultry factory farms that surround it. Beyond Factory Farming, a Canadian NGO, reported that compounds in hog manure, many of them toxic or pathogenic, may include hormones; human waste; hog carcasses; cleaning chemicals; insecticides; weed seeds; volatile organic compounds; salts; nitrogen; phosphorous; potassium; calcium; metals such as cadmium, zinc, nickel, lead, iron, boron, copper and manganese; vaccines; antibiotics; antibiotic-resistant bacteria; parasites; and other bacterial and viral pathogens such as *Cryptosporidium*, *Salmonella* and *E. coli*.[261]

The International Institute for Sustainable Development, among others, has proposed water quality trading for Lake Winnipeg. This would require setting overall targets for nutrient (pollution) reduction in the lake and its tributaries and then allowing water quality trading, so the big industrial farm operations on Lake Winnipeg could pay to continue to pollute.

Wall Street loves water

Water is big business. Investors are flocking to companies like Veolia and Thames Water that deliver private water services and wastewater treatment and water technology companies that clean contaminated water. Other money-making water investments include desalination projects and water pipeline construction. The bottled-water industry is booming, and investors are flocking to lucrative water markets where water is bought for industrial farming and mining or moved from its place of origin to be sold elsewhere. Wall Street's entry into water markets is a new cause for alarm, says Food & Water Watch, due to large funds and the unapologetic profit drive that investors bring to water markets. Some of the world's largest investment banks, such as Bank of America and UBS, are buying up farms and other lands in the American West with water claims.

Some water investments, such as building municipal water infrastructure and cleaning up old, contaminated sites, serve the public interest while they make a profit for the companies involved. This is a legitimate and important role for the private sector. But others take advantage of our collective and continued abuse, mismanagement and outright plunder of the planet's water heritage. If we had good public water services everywhere, there would be no need for private water utilities. If extractive industries were forced to clean up their practices, there would be far less damaged water. If we didn't still ruthlessly pollute ground and surface water all over the world, there would be no need for bottled water and far less need for the treatment of contaminated and wastewater. If we had proper laws protecting watersheds from harmful chemicals and toxins from factory farms and unregulated industry, there would be clean water for all. It is hard not to think that, for some in the water business, the absence of strong regulation and protection of source water is a ticket to windfall profits.

The lack of public funds for water services and source water protection is a major reason for the privatization and commodification of water and has created a huge opportunity for those in the water business. One might hope that the private sector is at least investing back into delivering water services and protecting source water. But according to a 2025 World Bank report, 91% of annual spending on water services globally comes from the public sector — inadequate as it is — with less than 2% contributed by the private sector.[262]

Global law firm White & Case writes that the lack of government funds, plus fragmentation of water delivery and resource management in Latin America, has "unlocked private investment in the water sector." It notes the raw sewage and pollution that pours into the increasingly damaged waterways of Latin America. It also points to the staggering number of people living without water and how the region will not be able to achieve the UN Sustainable Goals for water and sanitation without the influx of money it claims the private sector will bring. After years of fighting private water utilities, governments are now seeking them out. Brazil adopted a legal framework opening up the water market for public-private partnerships and the number of private concessions in that country quadrupled between 2019 and 2023.[263]

Hervé Thiard, managing director of the Swiss transnational private bank Pictet, told the French economics magazine *Capital* that profits in the water sector are growing by more than 6% a year, giving an annual return of 8%. "Water falls from the sky, so how can it be profitable? Because water is never available where we want it, in the quantity and quality we want. Less than 1% of the world's water is directly consumable; the rest is salty, frozen, polluted or simply far from where it is needed. It must therefore be collected, purified, transported, distributed, controlled and treated . . . in fact, water is an industry."[264]

Demand for water will soar as conflicts in the Middle East and other places intensify, data centres consume vast amounts of water resources and the global population grows, said Nacho Moreno, CEO of the Spanish utility group Cox, as he launched a US$328 million public offering on the Madrid Stock Exchange in October 2024. The water treatment business alone will double every four or five years, and there is money to be made. Moreno told the *Financial Post* that if Saudi Arabia's desalination plants were to blow up, the country would run out of water in a few hours. His company is developing floating desalination plants that can be moved around as needed. "We're going to grow on water."[265] As just one example of many, the water division of Abu Dhabi's utility company TAQA earned almost US$6 billion in 2024, largely on desalination and water pipeline projects.

Imagine a drying planet where desalination plants ring our oceans and coasts, converting massive amounts of seawater to drinkable fresh water that must then be pumped long distances to service farms, cities, businesses and homes. At the rate we are going, this scenario is not far-fetched. There are currently around 20,000 desalination plants in the world, found mainly in the Middle East and North Africa, with plans for many more. Desalination plants are energy intensive, highly polluting, very expensive — some cost over US$1 billion to construct — and notorious for their cost overruns. The brine they discharge back into the oceans contains heavy metals and chemicals and large quantities of concentrated salt that disrupt marine ecosystems and create dead zones where aquatic life struggles to survive. Environmentalists and scientists are sounding the alarm about the health of the Arabian Gulf due to the intense brine discharges from the region's many plants.

Yet because we are continuing to plunder the planet's fresh water sources, the desalination industry is set to soar. It saw substantial growth between 2018 and 2023, according to a January 2025 report by the Irish-based group Research and Markets and has a current market value of almost US$10 billion. The market is set to double by 2028 and continue to explode to a "robust" US$36.67 billion by 2033.[266] Pedro Arrojo-Agudo, the UN special rapporteur on the rights to safe drinking water and sanitation, has warned that desalination is not an excuse not to care for water and protect the water cycle. "But remember that when we talk about desalination plants, we must not think that since we have the sea here we can do anything, because then we will continue to create new spirals of unsustainability that will end up, at the very least, being very expensive. We cannot trust everything to desalination plants."[267]

Wastewater is another money-maker. India-based WiseGuy Reports estimated the 2023 wastewater market at over US$445 billion and expects it to grow to over US$1 trillion by 2032.[268] In 2025, industry news journal *Water Online* reported that the US wastewater market dealing just with forever chemicals PFAS is expected to grow at over 11% by 2031 and estimated it is now worth over US$90 million.[269]

All this is taking place with the approval and support of the international institutions guiding global policy on water. In March 2024, the World Bank and its private sector arms, the International Finance Corporation, the Multilateral Investment Guarantee Agency and the 2030 Water Resources Group that promotes public-private partnerships, created the Strategic Framework for Scaling Up Finance for Water to "catalyze financing and innovation for the water sector." The consortium calls on governments to establish "enabling conditions and reforms" where private capital can "develop a pipeline of bankable projects to attract this capital." They note, "The private sector is vital as an off-taker, financier, investor, and operator."[270]

In early 2025, the EU Commission called for public responses to its planned European Water Resilience Strategy to address the issues of aging water infrastructure and climate change in Europe. Many water activists and experts were concerned about the mission letter from the president of the commission to the commissioner for environment that said, "You will work to facilitate public and private investment in technology and cross-border water infrastructure." In its brief to the commission, the European

Public Service Union expressed strong concern about the "dangers of further liberalization" of water services and the promotion of public-private partnerships, which are "incompatible with the need to ensure universal access to water, a fundamental human right, in times of increasing water scarcity."[271]

The financialization of water has taken another step with the creation of water futures. In December 2020, the Chicago Mercantile Exchange, the world's largest financial derivatives exchange company, launched the world's first water futures market in California. Futures are forward-looking contracts where a buyer and a seller agree to exchange a commodity for a fixed price. The transaction is about the asset, as it does not involve the physical trading of actual commodities. If the price of the commodity — in this case, water — goes up, the buyer can make a profit. Understandably, there is a great deal of speculation involved.

Food & Water Watch reported in 2021 that WestWater Research, a water trading company, established an algorithm for estimating the cash price for the exchange of water allocations in California, and Veles Water Limited, a financial firm focused on water and NASDAQ, created the NQH2O Index based on the algorithm. The temptation for institutional investors to drive up the price of water entitlements is obvious, and Food & Water Watch warned that a large presence of "massive passive" speculators in the water market could send signals that the price of water will increase, leading to physical water hoarding and higher water prices. The impact would drive out small producers and further consolidate farmland into corporate businesses.[272]

Not surprisingly, there are now calls for a Global Water Pact based on the principles of the financialization of nature. In October 2024, the Global Commission on the Economics of Water, a project of the Organization for Economic Co-operation and Development and the UN, released a major report called *The Economics of Water: Valuing the Hydrological Cycle as a Global Common Good*. The report rightly sounds the alarm on the global water crisis, calling for a "sea change" in how we understand and act on water. It pays lip service to the human right to water and describes water as a "global common good," both positive in themselves. However, there is little doubt that the authors have a market approach to protecting the planet's water.

They call for putting a price on water, shaping markets "to spur a wave" of investments across the entire water cycle, forging "symbiotic partnerships between the public and private sectors" to deliver sustainable water use, developing financial institutions "to provide catalytic finance to unlock vastly greater amounts of private finance," creating a new global water data infrastructure bank, valuing water as "natural capital" to enable responsible stewardship of freshwater ecosystems, and measuring the benefits of green water (water held in soil and plants) to "enable schemes for Payment of Ecosystem Services." The same language used to shift the burden of protecting biodiversity from governments to the private sector clearly defines this as a financialization of water project.[273]

Ocean grabbing

A recent and growing field of private sector investment and profit is in Earth's oceans. Blue carbon refers to the carbon sequestered in coastal ecosystems, mangroves, seagrasses and salt marshes, large areas of which have been degraded or destroyed. Estimates of the total amount of carbon stored in these natural ecosystems globally are highly uncertain, reports Climate Analytics, a global climate science and policy institute dedicated to climate action.[274] They warn that using unreliable estimates to allow the use of carbon offsets may be a serious mistake. In fact, in a January 2025 report issued by the science journal *Global Change Biology*, scientists and researchers from the University of Southern Denmark found that blue carbon ecosystems sequester far less carbon than previously claimed, an overestimate by up to 18 times.[275]

But blue carbon trading has already started, and it works the same way as other nature-trading schemes. "A Blue Bond is a debt instrument that national governments, development banks and corporations issue to raise finance for marine and ocean-based projects that have long-term sustainability objectives and benefits. Blue Bonds work like conventional bonds wherein an investor lends capital to the bond issuer who pays back the initial investment plus an interest rate every year until the end of the bond's term," explains the European Commission. Blue bonds can be listed on stock exchanges, and long-term liability investments, such as pension funds, are attracted to them, it adds.[276]

The EC teamed up with United Nations Environment Programme, the World Wildlife Fund, the World Resources Institute and the European Investment Bank to create principles for the financial industry to direct investment in projects contributing to Sustainable Development Goal 14, on protecting the oceans. However, despite the noble-sounding nature of this goal, the reality is that blue bonds are just another way for investors to make a profit from "saving" nature while governments fail to protect marine ecosystems.

To see the real agenda, we must look to the market promoters. According to a report by Carbon Credits, a media platform promoting the carbon business, coastal habitats are a "75-gigaton carbon sink" and provide "a new opportunity" to generate blue carbon credits, "tradable on carbon markets for entities wanting to offset their GHG emissions." Carbon Credits cites mangrove projects between Apple and Conservation International in Colombia and Procter & Gamble and Conservation International in the Philippines. You can "make your money grow" buying blue credits, including equity or debt financing, exchange-traded funds and direct investment, the report says.[277]

Indonesia, with support of the UN, issued the first publicly offered sovereign blue bond credit in June 2023, raising US$150 million to scale innovative financing instruments and develop the capacity of stakeholders to meet the UN Social Development

Goals. The first blue bond in shipping, amounting to US$70.45 million, was issued in early 2024 by Japanese shipping heavyweight Mitsui O.S.K. Lines. In December 2024, Dubai-based DP World, a large port terminal and cargo operator, issued a "staggering" US$100 million blue bond, becoming the first company in the Middle East to mark this "milestone," reported *Offshore Energy Newsletter*. The company envisions this move as a path to link global trade within a "sustainable blue economy."[278]

Ocean pasture restoration (OPR) is a highly controversial form of geoengineering that uses iron fertilization to stimulate phytoplankton growth in the ocean with the intention of sequestering carbon and regenerating "ocean pastures." It was pioneered by American businessman Russ George, who founded the first OPR company, Planktos, in 2007, but the company failed to find funding for an experiment off the Galapagos Islands. Several experiments — one off the coast of Alaska and another off the coast of British Columbia — caused such concern over possible side effects that the UN placed a moratorium on what it called ocean fertilization.

But it is making a comeback as a form of carbon offsets for the private sector. OPR World Inc. is a London-based "privately financed public-private partnership" that promises to deliver large-scale ocean eco-restoration and such profits that will allow it to become a "major taxpayer."[279] The European Parliament is actively involved in ocean pasture restoration through the EU Nature Restoration Law, and recently adopted a multibillion-pound European Ocean Pact whose stated goals include "boosting the competitiveness of the EU sustainable blue economy."[280] While the stated goals of restoring the seas is, of course, vital, many fear that geoengineering corporations may be given a green light to experiment with something we know little about.

Pat Mooney is a Canadian researcher and writer considered one of the world's leading critics of geoengineering. He warns there have been no controlled experiments to prove that ocean pasture restoration will not create more problems than it is purported to solve. He particularly opposes commercial ocean fertilization schemes as an unacceptable attempt to profit from a global climate crisis. Greenpeace warned a gathering of the UN Convention of Biological Diversity years ago that ocean fertilization could prompt toxic tides and lifeless waters, disrupt ecosystems and have negative impacts on coastal livelihoods. The organization has maintained its strong opposition to ocean engineering and calls for the establishment of ocean sanctuaries to allow marine life to recover naturally.

Watchdog online journal *Geoengineering Monitor* expressed deep concern in 2023 over moves to allow ocean fertilization, warning that our oceans are becoming an experimental field:

> The geoengineering technologies that are being proposed to manipulate our ocean ecosystems and global natural processes come with great risks and uncertainties, as well as with foreseeable adverse impacts on marine ecosystems and human communities. These techniques are currently in their conceptual and early research phase. However, their effectiveness, feasibility and safety are often exaggerated in order to develop more funding and legitimacy for carbon dioxide removal research, often come with a heavy dose of exaggerated claims of efficacy, feasibility and safety — to say nothing of the risk that comes with offering today's main polluters — the fossil fuel industry — a perfect cover to continue polluting.[281]

The large global conservation organizations are often deeply involved in blue bond projects, promoting them in the first place and negotiating the agreements with governments, local communities, international financial institutions and investors. The World Wildlife Fund is strongly promoting "ocean conservation finance" in Canada, noting that Canada has the longest coastline in the world. In promotion sites replete with photos of driftwood, wild salmon and seagrasses, it calls blue bonds a form of "protection, stewardship, and restoration" and highlights the need to recognize Indigenous land rights and treaties in moving forward. WWF promotes carbon credits so that individuals, corporations and governments can "purchase carbon offsets to compensate for their emissions elsewhere."[282]

WWF helped develop an Indigenous-led conservation blue bond project in the Great Bear Rainforest, considered one of the biggest carbon sinks in the world. As described in a January 2024 CBC report, the carbon offsets are sold by Coastal First Nations to compensate for revenue lost from an agreement to reduce logging in the forest and to fund the Guardian Watchmen, Indigenous conservation rangers who patrol the coasts and forests. The offsets in this agreement are clearly understood to "cancel out greenhouse gases that are spewed into the atmosphere elsewhere."

Ian McAllister, a British Columbian writer, explorer, filmmaker and co-founder of the wildlife conservation organization Pacific Wild, says that outside the relatively small and protected area of the Great Bear Rainforest (about 30% of the region, but much of it is rock and ice), which does not allow for logging, there is ongoing and escalating logging. "The value of these trees is clearly worth more to cut down than any carbon scheme or we wouldn't be witnessing so much logging in the area. The entire premise of Ecosystem Based Management (EBM) that was meant to offset any destruction caused by logging was a total joke that has of course ended up being ignored. No independent monitoring or audits. No funding to enforce the basics of EBM. It has basically just disappeared and it is back to business as usual with logging, LNG, Fish Farms and the rest," he wrote in an email.

Even if the program were working for some local communities in the protected area, the simple fact is that this sacred forest is being used to allow corporations operating elsewhere to continue to pollute. But early reports show that the revenue from offsets has not met expectations, likely because the higher standards for accreditation and Indigenous oversight in British Columbia have made this project less competitive than some projects in the Global South that lack proper regulation.[283] As these blue bond markets increase, it is very likely we will see a competition for investors, selling the credits at rock-bottom prices. As it almost always does, the market will operate to challenge higher standards and regulations to turn a profit.

The Transnational Institute (TNI) is highly critical of blue bonds. In marketing them, the World Bank and other promoters call blue bonds a win-win-win on ecological, social and economic fronts. In practice, says TNI, the contradictions are glaring. The blue bond market is based on the "selling nature to save it" model that quenches the social thirst for action while turning conservation into a profitable venture that, in turn, accommodates destructive uses of ocean space. And just as forest and land-based carbon trading has not reduced overall emissions, there is no reason to trust that ocean-based carbon trading will either.[284]

In 2017, Climate Analytics warned that the use of blue bonds to offset and hence effectively avoid required emissions reductions in other sectors such as fossil fuel combustion, industry, agriculture, international aviation and marine activities would undermine our ability to limit temperature rise to 1.5°C. Claire Fyson, a scientist with the Potsdam Institute for Climate Impact Research, wrote, "There is very little room for offsetting in mitigation pathways compatible with the Paris Agreement due to the very limited carbon budget, yet there are already attempts to use the land sector and now the coastal ecosystems to contribute to national mitigation targets and reduce the emissions reductions needed in other sectors. This is particularly concerning for coastal ecosystems as carbon measurements are highly uncertain and carbon sequestration may, or is even likely to, be reversed by the increasing impacts of climate change."[285]

As well, blue bonds, also called "ocean conservation finance," feed the growing global demand for food and nutrition through the expansion of capital-intensive large-scale aquaculture, while brushing off its negative impacts on local populations. In the past 30 years, aquaculture production has exploded, outpacing the growth of capture fisheries every year, and now produces over half the fish we eat.

Mangroves provide essential protection for coastal communities throughout the world. They provide nursery grounds for fish and barriers to storms and hurricanes. But, says the UN Development Programme, at least one-quarter of the world's mangroves have been lost in the last 40 years despite all the claims of protection through carbon offset projects. The biggest culprit, according to a study by an international group of scientists with Global Mangrove Watch, is aquaculture. "Mangroves are being threatened across their entire range and have suffered large losses over the past century, primarily due to increasing coastal populations and the conversion of mangroves to aquaculture," said researcher Nathan Thomas.[286]

Beyond aquaculture, a great deal of the investment interest in marine ecosystems is less about protecting them than exploiting them for ecotourism and the search for rare-earth in the seabed. To gain international support for their plight facing rising seas, the Pacific Island States have adopted a new "clean-energy-based blue economy" framework, granting concessions for deep-sea mining to source the key elements for wind turbines and photovoltaic batteries. The Transnational Institute writes that the "blue fix" promoted by these projects is an attempt at accommodating the interests of capital in expanding accumulation opportunities through existing and new industries, while maintaining some form of social legitimacy by promising that these new uses do not undermine sustainability because they are combined with conservation: "Dressing up all these destructive practices in 'blue' language doesn't change the nature of their material socio-ecological impacts for people and the planet. If we really want to move towards a more sustainable use of coastal and marine resources, we have to fundamentally challenge the expansion of many, if not all, of these ocean industries that are currently shacking up with Transnational ENGOs — while pushing out former users of these ocean spaces, notably small-scale fisher peoples. Tragically, small-scale fisher peoples are, in most cases, the only actors that can truly be said to be using these ocean-spaces in a sustainable manner."[287]

Debt for blue nature swaps

Another issue with the blue market is the influence "debt-for-nature" swaps give foreign investors and asset managers to dictate policy in the countries where they are operating. US-based magazine *Jacobin* cited the case of the endangered Galapagos Islands. In May 2023, the government of Ecuador signed a financial debt-for-nature agreement with a collection of environmental NGOs, banks, law firms and investors — the largest of its kind in history — to "save" the Galapagos with promises of projects to be carried out by a Delaware-incorporated trust. The deal, backed by insurance from the Development Finance Corporation and a guarantee from the Inter-American Development Bank and structural advice from Credit Suisse, reduces the country's debt burden by over US$1 billion. In exchange, the country will take US$450 million of its debt service payments and direct it to conservation efforts in the Galapagos.

Critics say the yet-to-be-determined projects have no clear mandate and no oversight or enforcement mechanism to ensure that the funds will be used for conservation. One fear is the promotion of further ecotourism on the Islands. Former Ecuadorian environment minister Daniel Ortega Pacheco says the deal is essentially Ecuadorian money ostensibly being used for Ecuadorians, but where the administration has been ceded to private actors. *Jacobin* warned that the deal won't save the Galapagos. "In exchange for a reduction of the national debt, Ecuador has ceded sovereign control of the Galapagos Islands to an independent trust based in the

United States. The trust promises to invest sustainably — but no one can make it follow through."[288]

Africa is turning to private finance to fund its ocean climate plans, including the Great Blue Wall initiative that aims to protect coastal and marine areas across the continent's east coast from Somalia to South Africa and a number of islands, including Madagascar and Seychelles. The African Development Bank is seeking up to US$1.6 trillion to fulfill Africa's climate commitments and is looking to blue bonds to fund much of it. Transnational Institute says blue bonds are being heavily promoted in African countries as a way to meet their 30 x 30 goals, promote "blue growth" and create "blue jobs." In an in-depth 2023 study, the Transnational Institute noted these debt-for-ocean swaps manipulate the debts of poor countries, leading to a transfer of wealth and power to unaccountable US conservation organizations now working in close partnership with investment firms and banks.

TNI questioned the so-called "funding gap" — the widely held message that the issue in marine and biodiversity conservation is a lack of money. "A lack of finance has not depleted the oceans of fish or caused pollution and climate change. To believe this story is to fall into the trap of equating conservation with poverty. A more convincing view is that affluence and the pursuit of it have been the dominant theme in the destruction of nature." The "financialization of conservation" through blue bonds further entrenches the reckless view that saving nature must produce never-ending profits for the private sector.[289]

French environmental lawyer Raphaëlle Gautier warns that blue bonds are really about privatizing marine protection and threaten to silence the voices of marginalized peoples. The debt-for-ocean conservation swaps involve a complicated set of negotiations, involving governments, international finance institutions, banks, global conservation organizations, investors and many lawyers and accountants. Usually there is some urgency to the process as the countries are under pressure to meet the loan requirements to avoid increased payments in case they fail to meet the milestones of the deal. The right to free, prior and informed consent of local Indigenous people, peasants and fishermen is often hurried and inadequate, their voices drowned out.[290]

Economist and philosopher Adam Smith wrote in *The Wealth of Nations*, "As soon as the land of any country has all become private property, the landlords, like all other men, love to reap where they never sowed, and demand a rent even for its natural produce."[291] These words apply to our lakes and rivers and oceans as well as land. The privatization of fresh water and oceans to "save" them is a grave mistake.

CHAPTER FOUR

OUR RIGHTS-BASED FUTURE

We are not protecting nature. We are nature protecting itself.

— CASEY CAMP-HORINEK,

ACTIVIST AND PONCA NATION ELDER

Rita Mestokosho is an activist, poet and educator from the Innu community of Ekuanitshit located on the Gulf of St. Lawrence in Quebec. Her people are the guardians of the first river in Canada to be given legal rights. The Magpie, or Mutehekau Shipu to the Innu, is a magnificent river famous for its pristine forests, whitewater rafting and rich wildlife, that travels almost 300 kilometres before emptying into the Gulf of St. Lawrence. But Mestokosho's people have been very worried that Hydro-Québec, the fourth largest hydro-power producer in the world, has plans to further dam the Magpie for hydro exports, as they have most of the province's most powerful rivers to date. They are only one of many communities in Northern and Eastern Quebec worried that the growing appetite for Canadian hydro exports to the US will spur the building of more and more dams on their once free-flowing rivers.

As reported by *Canadian Geographic*, the local people saw the damage caused by a massive four-dam project on the neighbouring Romaine River and decided to act. Working with environmental organizations such as the Montreal-based International Observatory on Nature's Rights and SNAP Québec, the Innu and municipal councils formed an alliance to protect the river by giving it personhood status and assigning stewardship to Indigenous guardians who monitor and speak for the Magpie. In February 2021, the river was granted nine "rights," including the right to flow, the right to maintain its natural biodiversity and the right to be free from pollution. It even has the right to sue!

The Magpie declaration is part of a growing international movement to protect rivers and nature from destruction. Whether this declaration could stand up in a court of law is, of course, unknown. And given that most of these projects worldwide are led by Indigenous communities, their growing rights as laid out in the UN Declaration on the Rights of Indigenous Peoples, with its promise of "free, prior and informed consent," stand to be tested. Canadian human rights lawyer and former UN special rapporteur on human rights and the environment David Boyd says that Hydro-Québec would be taking a great risk if they think they can proceed with a dam under this law, as Canada's Constitution protects "Aboriginal Rights."

Boyd welcomes these rights of river challenges in Canada and internationally, as he says the Indigenous worldview is key to protecting the Earth. "Nature is far more than just a basket of resources; it's an extraordinary, wondrous community that we're incredibly fortunate to be part of. All Canadians need to rethink our fundamental relationship with nature if we are going to achieve a sustainable future," he said to *Canadian Geographic*.

"We have always known the river is alive, our ancestors have always said that," Mestokosho told *Canadian Geographic*. "The river is like the blood that runs in our veins. If the river is sick, we will also be sick. That's why we need to protect her." Her niece Lydia invites those who would dam the river to come and see it. "Do you really want to be the person who destroyed ancient portages and animals and their habitat for money? They don't understand the vision that the river is alive. But how can you see a corporation as a person, an entity with moral rights, and not see a river in the same way?"[292]

You may be asking, What does it matter if asset managers, banks and corporations have recognized the serious environmental crises we all face and taken up the call to "go green"? Isn't it better to have the private sector inside the climate tent helping to find a solution than outside pretending it doesn't exist? (We are speaking here about large companies with money to invest globally, not local businesses that are often community and environmental leaders.)

Without regulations, private companies are free to opt out of "green" promises

The first answer to this question is that we have to doubt the true commitment of many of these "green" corporate champions in light of how fast they flee when they no longer have to pretend to care. While there are no doubt individuals within these large private corporations and predatory investment firms who truly care about the environment, the very nature of global capitalism requires an adherence to the bottom line first and foremost. There are very few global companies that can successfully compete while paying to clean up their own environmental footprint unless legislation says they *all* have to. As proof, we need only look to how many large American and other transnational companies bailed on their green promises once Donald Trump became president.

In January 2025, BlackRock pulled out of the 2021 UN-sponsored sustainability initiative called Net Zero Asset Managers. Dozens of others followed. "Bill Gates Gives Up on Climate Change," blared a March 2025 headline, noting the closure of Breakthrough Energy, his non-profit policy and advocacy office on climate change. Oil giant BP, which had set some of the most ambitious targets to cut oil and gas production and ramp up investment in renewables, has "reset" its goals back to traditional oil and gas production after seeing how those targets had made it uncompetitive. Jeff Bezos's US$10 billion Earth Fund halted its support of one of the world's most important carbon certification organizations, the Science Based Targets Initiative. Professor Doreen Stabinsky of the College of the Atlantic, who is also a member of the initiative, told *The Guardian*, "Bezos is bowing down to Trump in a way a bunch of billionaires are bowing down to Trump."[293]

No sooner was Donald Trump elected president than six big US banks quit the UN-sponsored Net-Zero Banking Alliance, in an attempt to head off "anti-woke" attacks they assumed were coming. They were joined by six Canadian banks: Royal Bank of Canada, Toronto-Dominion, Bank of Nova Scotia, Bank of Montreal, National Bank of Canada and Canadian Imperial Bank of Commerce. In an April 2025 report, *The Guardian* wrote that some of these same banks now say it's time to face the fact that the world has failed to stop global warming. Recent reports by Morgan Stanley, JPMorganChase and the Institute of International Finance make it clear the finance sector considers the Paris climate agreement dead and is telling investors to plan accordingly. They do not sugar-coat what this means, saying the level of global heating will lead to heatwaves, floods, economic strife and other upheavals. In September 2025, Google scrubbed its net-zero pledge from its website after having positioned itself as a sustainability leader.

"We are clearly seeing a broad retreat on climate from the finance sector," Patrick McCully, senior analyst at Reclaim Finance, a group that pushes financial firms to act on the climate crisis, told *The Guardian*. "It is to a very large extent being influenced by Trump and his agenda of accelerating climate change, although also due to banks using Trumpism as an excuse to roll back commitments that they had never actually intended to keep." Never mind, says a Morgan Stanley report, the multiplying heatwaves will provide a windfall for companies that provide air conditioning, creating a market that could grow by 41% and be worth US$331 billion by 2030.[294]

There is no doubt people in positions of power in some of these companies see the dangers of the climate, water and biodiversity crises, even if just from the point of view of how they may negatively affect their businesses. But until there are rules that apply to all, companies that make serious commitments to environmental and social justice are at an economic disadvantage. Martin Luther King Jr. said, "It may be true that the law cannot change the heart, but it can restrain the heartless."

We have lost decades in the climate fight to this false solution

The second answer to the question as to why we should be concerned about the ascendency of a market "answer" to the climate crisis is that, in adopting it, we have collectively put off the desperately needed changes we must make if we and the planet are to survive. Simply put, the financialization of nature is failing before our eyes.

Nearly three-quarters of Earth's land is now degraded, and an estimated one million species face extinction. Earth's land masses are drying fast, notes the UN, pointing out that, in the last three decades, over three-quarters of all the world's land became drier than it had been in the previous 30 years.[295] Yet to keep pace with the global demand for food, crop production needs to grow by at least 50% by 2050, says the UN.[296] However, George Monbiot reminds us, to match crop production to this demand, water use for irrigation would have to increase by 146%. "One minor problem," he wrote in 2024. "Water is already maxed out."[297]

Groundwater levels around the world are dropping quickly, and climate change is warming groundwater as well as surface water, affecting millions. Scientists at the University of Reading report that the world's oceans are warming four times faster than they were in the late 1980s, and that for 450 straight days in 2023 and 2024, ocean temperatures hit record highs. Without substantial moves to cut fossil fuel use, they say, ocean temperatures will continue to rise "by a significant margin."[298]

The World Meteorological Organization reported in 2025 that the primary driver of our planetary climate crisis, CO_2, continues to pile up in our atmosphere at an accelerating rate — 2024 and 2023 were "record-busting."[299] The World Economic Forum reported in 2018 that air pollution is now more deadly than war, smoking and TB together.[300] All of these facts are known to political and corporate leaders around the world. Yet, according to *Carbon Brief*, nearly 95% of countries missed the February 2025 UN deadline to submit their new climate pledges for 2035, as the planet is set to exceed the Paris Agreement's target of limiting global temperatures to 1.5°C above pre-industrial levels.[301]

In a stunning piece of irony considering how many in the corporate sector have positioned themselves to be the saviours of the environment, a top insurer has warned that the climate crisis is on track to destroy capitalism. Günther Thallinger, on the board of Germany's Allianz SE, one of the world's biggest insurance companies, has warned the world is fast approaching temperature levels where insurers will no longer be able to offer cover for many climate risks. He told *The Guardian* that at the rate carbon emissions are rising, we are assured of rapid global temperature increases. At 3°C, the damage will be so great, governments will be unable to provide financial bailouts and insurance companies will no longer be able to offer coverage for many sectors. Thallinger explained,

> Heat and water destroy capital. Flooded homes lose value. Overheated cities become uninhabitable. Entire asset classes are degrading in real time. . . . This applies not only to housing, but to infrastructure, transportation, agriculture, and industry. . . . The economic value of entire regions — coastal, arid, wildfire-prone — will begin to vanish from financial ledgers. Markets will reprice, rapidly and brutally. This is what a climate-driven market failure looks like. . . . That means no more mortgages, no new real estate development,

> no long-term investment, no financial stability. The financial sector, as we know it, ceases to function. And with it, capitalism as we know it ceases to be viable.[302]

Urgent collective action is needed now. Waiting for the market to save the planet is pure folly.

While inequality and injustice have deepened

Our trust that the private sector can save us from climate collapse is problematic in another way: it has not only *not* challenged the deep inequality and injustice in our world, it has been a major driver. In the last decade of green finance, the wealth gap among and within nations has grown dramatically.

In its 2025 report, *Takers Not Makers*, Oxfam says that in 2023, the top 1% in rich countries gained US$30 million an hour. Billionaire wealth rose three times faster in 2024 than the year before, and 204 new billionaires were created. If any of the 10 richest men lost 99% of their wealth, they would still be a billionaire. The organization predicts at least five people will become trillionaires within a decade, and they will likely include Tesla's Elon Musk, Amazon's Jeff Bezos, Oracle's Larry Ellison, Meta's Mark Zuckerberg and LVMH's Bernard Arnault.[303] Many of them are supporters of Donald Trump. Bezos, Zuckerberg and Musk attended Trump's inauguration. Musk alone poured US$277 million into Trump's re-election campaign.[304] After his breakup with Trump, Musk posted on X, "Without me, Trump would have lost the election."[305]

Meanwhile, the number of people living in poverty has barely changed in 35 years. "People living in poverty all over the world continue to face multiple crises — from the scars of the pandemic to conflict and climate breakdown, driving further poverty, hunger, and inequality. Inequality is out of control," says Oxfam.[306] Save the Children reported that in 2024, the year so many new billionaires were created, at least 18.2 million children — 35 a minute — were born into hunger.[307]

In *A World of Debt 2024*, the UN Conference on Trade and Development reported that global public debt reached a record high of US$97 trillion in 2023 and that it is growing twice as fast in developing countries as developed. Small wonder that Oxfam reported in 2023 that the world's poorest countries would have to slash public spending by 57% — US$229 billion — between 2023 and 2027. Low- and lower-middle-income countries are forced to do this as they have to pay nearly half a billion dollars every day in interest and debt repayments. Entire countries are facing bankruptcy.[308]

The wealth divide exists within countries as well as between them. There is a "Third World" in the "First World" and vice versa. The top 20% of Canadians hold more than two-thirds of the country's wealth.[309] The richest 1% of Indians own 58% of the wealth, while the richest 10% own 80%.[310] The top 10% in South Africa hold as much as 85% of that country's wealth.[311] A May 2025 study published by the journal *Nature Climate Change* clearly linked the wealth gap to the heating of the planet, reporting that two-thirds of global heating is caused by the richest 10% of the world.[312]

In *The Value of a Whale*, Adrienne Buller calls the concentration of wealth and power the "bread and butter" of climate and ecological breakdown. She points out that as global unemployment and inequality hit an "eye-watering" scale during the Covid pandemic, inflows to exchange-traded funds surpassed an all-time record. "Thus from the perspective of the asset management industry, wealth inequality both within and between nations is optimal, as it provides a constant stream of individuals with savings to maximize by betting on stocks, bonds and real estate," she writes.

As the poor get poorer, they will sell whatever they can to survive and the growth of these billionaires means there are plenty of buyers. And debt-for-nature swaps have left many countries of the Global South even more vulnerable. In search of "cheap nature" to exploit — which exists in abundance in poor countries and on

Indigenous and rural lands in rich countries — these private concentrations of wealth continue to drive completely untenable rates of resource consumption and waste, rising emissions and dispossession of the global poor.

Even in the search for energy alternatives, such as electric cars and carbon capture and storage, there is profit to be made from the cheap labour and lax environmental rules in many countries. "Walls are erected and barbed-wire fences extended to exclude certain people from the future that green capital envisions. Swathes of the planet are designated as the necessary sacrifice zones of achieving that future," Buller writes.[313]

"Sacrifice zones" exist in the Global North as well. In the US, whole low-income, less-educated and minority communities were hollowed out when manufacturing jobs moved to Mexico and Asia. Low-income and minority communities also disproportionally bear the environmental brunt of power plants, factory farms and chemical industries that are often situated near them. Food & Water Watch reports that pollution trading compounds environmental racism and that market-based programs like cap-and-trade compound the harm to vulnerable communities.[314]

In her 2025 book, *Climate Injustice: Why We Need to Fight Global Inequality to Combat Climate Change*, climate scientist Friederike Otto, with Imperial College in London, asserts that the climate crisis is a symptom of global inequality, injustice and social vulnerability. She told *The Guardian* that those who benefit from the burning of fossil fuels are the few wealthy people who have stakes in or own the companies themselves. The majority do not benefit. "What turns weather into a disaster is not how much it rained but how vulnerable people are and how well prepared. Therefore, depending on which type of weather event we are looking at and where in the world we are, we always find that the more unequal the society is, be that a US city or a state in western Africa, the more severe the consequence."[315]

The path to protect nature, restore biodiversity and fight for social justice in these chaotic times is not an easy one. But this much is clear: the deep connection between environmental destruction and social injustice and inequality demands solutions that address the two together. The rights of people and the rights of the Earth are inextricably entwined.

The way forward

Reject the financialization of nature

The financialization of nature is the wrong solution to the environmental and social crises we face. It must be rejected. The chorus of voices from the Global South calling for an end to the false solution of green finance is growing louder and must be heard. Those in a position of influence in the Global North need to listen. The fairly recent emphasis on protecting and restoring biodiversity at the highest levels of government, academia and international institutions is welcome. However, we have been slow to recognize that allowing markets to control the process has dazzled many into thinking we are adequately tackling the climate and biodiversity crises when, in fact, we are not.

There is a place for the private sector in helping to find shared solutions, from creating innovative new carbon capture technologies and building energy-efficient homes and businesses to laying down the pipes and infrastructure for the public water systems so desperately needed all over the world. Also very welcome are commitments and measures taken by industries and companies to minimize their own environmental footprint in real ways, cutting their own carbon emissions and water pollution for instance, and investing in eco-friendly technology in their own operations.

There is a fundamental difference, however, between business being part of solutions created through democratic processes and overriding governments to set climate policy that enriches corporations and the wealthy and leaves most of the rest of the world at risk. The energy, mining, chemical and industrial agriculture corporations still plundering the planet in the name of nature protection must be exposed and opposed.

We need a serious debate about green finance and how so many of us have allowed market-driven solutions to what are life-threatening crises. One key reason that many environmentalists and climate change activists and organizations have been slow to critique or even question nature-based solutions is that this model appears to come from a shared concern and analysis. Putting nature at the heart of all we do, all policy, all economic decisions — as the green economy was supposed to do — is dear to the heart of all who care about the Earth. But it is urgent that climate, environmental and conservation organizations take a deep look at what it has really meant to our movement to trust powerful private forces to set the agenda and not question the price of offsetting the bad practices of large corporations onto poor and vulnerable communities.

This will mean challenging the support that corporations and asset managers are still getting from wealthy governments and institutions like the UN and the World Bank to promote a market framework in the name of fighting climate change. Greenpeace International calls on the European Union to stop the cash flow to businesses destroying nature. Six of the nine planetary boundaries that Earth scientists have defined as needing to be respected to ensure a stable global environment have already been breached, it noted in a March 2024 report called *Bankrolling Ecosystem Destruction*. Yet the EU spends billions funding industries and projects that accelerate environmental degradation and species extinction. Greenpeace calls for the drastic reform of a financial system that subsidizes the expansion of economic activities that lead to nature's destruction.[316]

Powerful voices from the Global South, including Global Forest Coalition, Third World Network and Friends of the Earth

International, call on the world to reject the commodification of nature through the monetary valuation of ecosystem functions and the creation of biodiversity markets:

> We call on governments, multilateral bodies, conservation organizations and other actors to prioritize transformational change in tackling the underlying causes of biodiversity loss, including: promoting effective regulation of harmful corporate activity; recognizing and respecting, protecting and promoting the right to land of Indigenous Peoples, local communities, small-scale food producers and women; stopping financial flows and investments that are harmful to biodiversity and peoples; removing harmful government subsidies; changing production and consumption patterns, especially of the rich; supporting a just transition; . . . pursuing effective and equitable means of conservation; and taking immediate steps to phase down the supply and use of fossil fuels.[317]

Adrienne Buller writes,

> Rather than accept the emerging capitalist hellscape of government incentives for private profiteering from "nature-based solutions," we need to articulate an alternative vision of why nature matters. We should demand that the protection and restoration of ecosystems and biodiversity prioritizes environmental and collective social wellbeing over financial returns — both domestically and internationally. Instead of "nature-based" offsets and land-grabs, we need reparations for our outsourcing of environmental destruction. In place of private enclosure, we need an expansion and protection of shared spaces, and to

> reclaim land for the commons. And we need an economy in which everyone has both enough time for and proximity to nature, allowing us all to reconnect to it as a source of wellbeing and wonder.[318]

Confront global inequality

Wild discrepancy in wealth renders large swaths of humanity without a voice to protect themselves or their natural surroundings from plunder. Nnimmo Bassey, Nigerian environmentalist, poet and former chair of Friends of the Earth International, said in an email exchange, "An ongoing grabbing of the African continent is predicated on the notion that making money from nature's capital is such a cool and ecologically sound idea. Unequal ecological exchange continues!" To confront this unequal exchange means confronting not only the deep and continuing exploitation of the Global South, but also the deep and continuing wealth inequality that prevents countries from caring for their people or their environment.

Oxfam notes that 2025 was the 70th anniversary of the Bandung Conference, where Global South countries sought to create a new international order to oppose colonialism and neocolonialism. It led to the Non-Aligned Movement — a forum of 120 countries to advance the interests of the developing world. Alas, reports Oxfam, things are not much better today. Modern multinational corporations are a creation of colonialism and continue to exploit workers and extract the wealth of the Global South, "capturing the vast majority of profits and perpetuating dependence, exploitation and control through economic measures," it says in *Takers Not Makers*.[319]

Oxfam calls for a realignment of the global system through meaningful systemic change. Set global and national goals to end extreme wealth so that the incomes of the richest 10% are no higher than the poorest 40% globally. Repair the wounds of historical colonialism by acknowledging it and setting the stage for reparations. Change the governance of the World Bank, the International Monetary Fund and the UN to end the dominance of the Global

North and the interests of their wealthy and elites. Tax the richest to end extreme wealth with a global tax policy convention to facilitate the payment of higher taxes by the rich and radically reduce inequality. Strengthen public services and implement land reform to ensure access to land and housing. Finally, says Oxfam, promote North-South cooperation and solidarity by fostering collective development of shared knowledge, technology and resources to further sustainable development.

Calls are also growing for debt relief. The Global South is on the brink of a disastrous debt crisis, say leading development economists. Indeed, according to *Development Finance International*, citizens of the Global South now face the worst debt crisis since global records began. Low-income and middle-income countries, which have seen the amount paid on foreign debt payment increase by 150% since 2011, are being hit especially hard, reported *Truthout* in 2024. "A debt crisis of epic proportions is emerging in the Global South. Some have referred to this as a 'silent debt crisis.' But it's loud and clear. We are poised on the cusp of a new 'lost decade' with vast debt overhangs, widespread debt distress, demands for austerity by lenders, and severe economic slowdowns, just some of the legacies of this crisis," said Ilene Grabel, professor of international finance at the University of Denver.[320] Serious debt relief goals are crucial to prevent further human and environmental abuses of the Global South.

Hundreds of organizations have signed the Manifesto for an Ecosocial Energy Translation from the People of the South, which clearly links poverty, hunger, poor health care and the lack of water in the Global South to the new "green" manifestations of colonialism. They decry the "green energy transition" whereby wealthy countries of the North are trying to decarbonize their own economies by securing critical raw materials found in the South. Poor countries are forced to "yield up" cobalt and lithium for high-tech batteries, balsa wood for wind turbines, land for large solar arrays and new infrastructure for hydrogen megaprojects.

"This decarbonization of the rich, which is market-based and export-oriented, depends on a new phase of environmental despoliation of the Global South, which affects the lives of millions of women, men, and children, not to mention non-human life. In this way, the Global South has once again become a zone of sacrifice, a basket of purportedly inexhaustible resources for the countries of the North." The manifesto calls on countries and peoples of the South to work with allies in the North to reject the "false solutions" of this green energy transition and advance a just and popular ecosystem transition, including the payment of ecological debt. It rejects land grabs for solar and wind farms and the indiscriminate mining of critical minerals and demands the genuine protection of environmental and human rights defenders. It also calls for a halt to international trade and investment agreements that penalize countries that want to curb fossil fuel extraction.

"We still have time to start a just and democratic transition," urges the manifesto. "We can transition away from the neoliberal economic system in a direction that sustains life, combines social justice with environmental justice, brings together egalitarian and democratic values with a resilient, holistic social policy, and restores an ecological balance necessary for a healthy planet. But for that we need more political imagination and more utopian visions of another society that is socially just and respects our planetary common house."[321]

Confront social inequality

A rebalancing of global wealth and power also requires tackling the extremities of wealth within countries of the Global North and held by the elite of the Global South. Many voices are calling on governments to take immediate measures to increase taxation on the richest, including permanent increases to tax on their incomes and capital, one-off taxes to end crisis profiteering, and taxes on wealth that are high enough to significantly reduce inequality. Combatting tax avoidance, banning the use of offshore tax havens and restricting foreign ownership of essential services are key to taking back democratic control of the economy.

Speaking to a March 2024 NGO-organized conference on tax justice in Paris, bestselling economist Thomas Piketty says Europe made a grave mistake in 2024 by voting, along with the US, against the creation of a framework convention on fair taxation at the UN and calls for a renewed pact.[322] Renowned American economist Paul Krugman has long called for taxation on wealthy individuals and corporations, arguing that the US economy grew faster during periods when taxes at the top were relatively high than it did when they were low. "Tax the rich, help America's children," he wrote in 2021, noting that helping children, in addition to being the right thing to do, has big economic payoffs in the creation of a healthy and educated population.[323]

The call for economic justice comes from all levels of society. Patriotic Millionaires, a powerful group of wealthy Americans whose motto is "tax the rich, pay the people, and spread the power," calls for an Anti-Oligarchy Act to create a more equal society. Members would impose a tax on the wealth of the richest Americans "sufficient to reduce their wealth to a level in harmony with the ideals of democracy, amending the United States Constitution if necessary." They would also triple the minimum wage, exempt lower-income families from paying income tax, subject millionaires to a surtax on their taxes due, tax profits from buying and selling stocks and other assets, and tax inheritance over US$1 million.[324]

An August 2024 report by Britain's Tax Justice Network found that countries can raise US$2.1 trillion a year by following the example of Spain's successful wealth tax on the 0.5% richest households — double the funding needed annually for developing countries' external climate commitment. Calling this tax "feather-light" as it only applies to the richest households at a rate of 1.7% to 3.5%, the report opens the door to even higher taxes on the wealthy through international cooperation. The report says the root of the problem is the two-tier treatment of earned wealth and collected wealth, the latter typically taxed at far lower rates than the former. Half the wealth created around the world each year goes to people who earn for a living; the rest is collected in investments, interest, dividends and capital gains. Not only is this wealth not taxed fairly, it is less economically productive as much of it is diverted toward speculative derivatives instead of providing goods and services in the real economy.

"There's this idea that billionaires earn wealth like everybody else, they're just better at it. This is bogus. It's impossible to earn a billion dollars," said Mark Bou Mansour of the network. "The average US worker would have to work for a stretch of time 13 times longer than humans have existed to earn as much as wealth as the world's richest man has today. Salaries don't make billionaires, dividends and rent money do. . . . By definition, a billionaire owns more wealth than an average US household could spend in 10,000 years. Wealth contributes a lot less to the economy than it can when it's pharaoh-tombed like this, making economies poorer than the sum of their parts. To make our economies secure and protect the earner way of life that has defined the modern era, we need wealth taxes that end the two-tier treatment of wealth."[325]

Heed Indigenous voices

Indigenous organizations and communities have been on the frontline in opposing the commodification and financialization of nature, and they are often the first victims when abuses occur. With more than 28% of global land under some form of Indigenous management or tenure, their perspective is crucial. Indigenous people view themselves as part of the ecosystem rather than apart from it and understand they are the guardians of nature and biodiversity in their territories. They are sounding the alarm as their lands, forests and waters are being divided up to be "protected" by investors trading in carbon and water markets.

The US-based Indigenous Environmental Network asserts that the financialization of nature is a form of "greenwashing" that does not address the root causes of climate change and co-opts effective ecological practices based on traditional Indigenous knowledge. "The narrative and framework of nature-based solutions transforms effective ecological practices into financialized instruments that exacerbate the climate and biodiversity crises," said Tom Goldtooth, the organization's founder. Green finance is a public-relations scheme that "commodifies the sacred" and entrenches the exploitative power dynamics of colonialism. "The hidden agenda behind nature-based solutions is to facilitate the absorption of climate change policy further into the private sector."[326]

At the Climate COP25 in 2019, a new network, the Climate, Land, Ambition and Rights Alliance, reported that Indigenous and farmers' movements were going to reject the phrase "nature-based solutions" because it is "becoming a meaningless term that legitimizes harmful approaches." Instead, said a representative of ActionAid International, "We will be using the specific terms that we want to advocate for, such as restoration of biodiverse ecosystems, agroecology, etc., under a rights-based approach." Network members expressed strong concerns that concentrating on market-based nature solutions could serve as a smokescreen to hide the lack of progress in putting a stop to the use of fossil fuels to meet the goals of the Paris Agreement.[327]

Greenpeace says it's easy to understand why Indigenous Peoples are regarded as the guardians of biodiversity. They courageously resist the destructive plans pursued by governments and corporations interested only in extracting wealth from the Earth to line their pockets. "Indigenous Peoples are the stewards of their lands, which play a key role in tackling the climate crisis. In Brazil, studies show that between 2005 and 2012, deforestation rates were 17 times lower in Indigenous territories than in unprotected areas of the Amazon. Forests play a crucial role in storing carbon underground, and protecting them means protecting us from further escalation of the climate crisis."[328]

In 2007, the General Assembly adopted a landmark agreement called the United Nations Declaration on the Rights of Indigenous

Peoples. It set standards for the survival, dignity and well-being of Indigenous Peoples, affirming their rights to free, prior and informed consent before any project involving their land, territories or resources is approved. This declaration broke new ground in the understanding of human rights as it included in its right of self-determination, political, legal, economic, social, cultural and spiritual institutions; traditional knowledge; dignity and well-being; and conservation and protection of natural resources on Indigenous territory.

While not legally binding, this declaration has encouraged some key developments asserting the stewardship of Indigenous Peoples around the world to stop the plunder of their territories. In 2025, the Siekopai — a small ancient Indigenous nation that calls itself the "people of many colours" — was slated to return to its ancestral territory in the Amazon rainforest bordering Peru and Ecuador for the first time in 80 years, having been displaced first by slavery, then by war. With the support of several NGOs such as Amazon Frontlines, the Siekopai successfully challenged the legality of existing laws that allow for extractive industry on their land, and the courts granted them title to over 600,000 acres of forests and lagoons. As reported by *Wired*, they seek to expand their land title within Ecuador to protect 5 million acres in the Amazon and dismantle the legal and political barriers to their title of an estimated 40 million acres of ancestral Indigenous territory in Peru.

The Amazon is nearing a tipping point from which it may never recover. Its best chance is the stewardship of its first peoples. *Wired* explained,

> In 2025, the Siekopai will finally go home, protecting their culture, the headwaters of the Amazon, and ensuring that millions of trees keep absorbing carbon dioxide and producing rain. The Siekopai will also show the world that solutions to the climate crisis must focus on these displaced communities.

> The technological fixes of paper parks, carbon offset markets, and bureaucratic soil classifications are not working. And the irony is, indigenous communities that want to go home and protect their forests produce better carbon stats than the technocrats. They know how to leave the carbon in the trees, leave the oil under the soil, and leave the gold dust in the mountains. They deserve permanent ownership of their lands — for their benefit and that of the planet.[329]

Canada has been blessed with massive forests. Its boreal forest that stretches from Newfoundland and Labrador on the Atlantic Coast to northern British Columbia and the Yukon Territory is the largest intact forest left in the world. In 2022, the Natural Resources Defense Council reported its trees, soils and peatlands store twice as much carbon as the world's oil reserves and are home to more than 600 Indigenous communities. However, industrial logging is threatening this heritage, as each year, one million acres of intact forest are logged.[330] So relentless is this assault that, between this logging and wildfire exacerbated by poor forest management, the boreal forest is now releasing more carbon into the atmosphere than it stores.[331]

First Nations, Inuit and Métis communities have been at the forefront of the fight to protect Canada's forests, and their legal challenges derived from their land and treaty rights are a formidable bulwark against the plunder of this heritage. In 2014, in a landmark case, the Supreme Court of Canada ruled in favour of the Tsilhqot'in Nation in its dispute with British Columbia over commercial logging in their territory, recognizing their title to over 1,700 square kilometres of land. This was the first declaration of Aboriginal title in Canadian history and confirmed that the doctrine of terra nullius (that no one owned the land prior to Europeans asserting sovereignty) has never applied to Canada.

While they do not wipe out the years of colonial abuse, recent rulings and treaties have moved the bar. In early 2024, a landmark agreement was signed by the Seal River Watershed Alliance and the federal and provincial governments to designate a pristine watershed in northern Manitoba the size of Nova Scotia an Indigenous-protected and conserved area. The alliance is made up of Dene, Cree and Inuit guardians of this newly protected watershed.

That same year, the Inuit of Labrador signed a memorandum of understanding with the federal government to create a new Inuit Protected Area of up to 17,000 square kilometres. The Government of Canada describes the area as "home to polar bears, whales, dolphins, seals, breeding and migrating seabirds, waterfowl, and a variety of fish species. Ranging from highly scenic fjords to long beaches and mudflats, the area is a cultural and ecological treasure." It adds, "The proposed Inuit Protected Area/national marine conservation area initiative recognizes and respects Inuit stewardship, emphasizes the importance of Inuit Knowledge in conservation planning and implementation, and further supports the Nunatsiavut Government's Imappivut Marine Plan that safeguards the marine environment and Inuit interests, rights, and priorities. This ensures that the protection of the marine environment aligns with the cultural, spiritual, and economic values of the Inuit people."[332]

In a moving February 2025 ceremony, the Government of Canada recognized the Aboriginal title of the Haida Nation, seafaring people who inhabit an enchanting string of over 200 islands totalling about a million hectares off the northern coast of British Columbia. Rich in ancient forests and abundant fish, whales and kelp forests, Haida Gwaii has been the scene of disputes over who should have access to and authority over its natural bounty. In signing the agreement, then Prime Minister Justin Trudeau said the agreement grants inherent legal rights to the land and enshrines the right of the Haida to control their own destiny. "Haida Gwaii belongs to you," he said, adding that the landmark agreement "recognizes that the Haida people have lived here since time immemorial. That the Haida people have an inalienable right to use, manage and enjoy the lands of Haida Gwaii as they see fit. That self-determination is the only path toward true reconciliation."[333]

Self-determination and Indigenous rights are also crucial to presenting a different worldview of nature and rights than that offered by the market and are key to fighting the financialization of nature.

Support the rights of nature

In 2010, following the spectacular failure of the 2009 Copenhagen climate summit, thousands of activists from around the world gathered in Cochabamba, Bolivia, to alter the discourse on how to deal with the climate crisis. Out of this historic meeting came the Universal Declaration on the Rights of Mother Earth, recognizing that Earth is an indivisible living community of interrelated and interdependent beings with inherent rights. One of the lead authors was Cormac Cullinan, a South African environmental and human rights lawyer whose seminal work, *Wild Law*, helped launch the rights of nature movement.

Cullinan believes that future generations will look back on ours and view our relationship with nature as a form of slavery. He explains that most legal systems view nature as property and that most laws to protect the environment and other species merely regulate the amount of damage that can be inflicted by human activity. He calls for laws that regulate humans in a manner that allows other species to fulfill their evolutionary role on the planet. Human laws and governance systems, he writes, must promote human behaviour that contributes to the health and integrity not only of human society, but also of the "wider ecological community."[334]

In his groundbreaking 2017 book, *The Rights of Nature: A Legal Revolution That Could Save the World*, David Boyd says that protecting the environment is impossible if we continue to assert human superiority, universal ownership of all land and wildlife and pursuit of economic growth. He cites the 1982 UN World Charter for Nature that clearly stated that every form of life is unique, "warranting respect regardless of its worth to humans." He also quotes a 1991 report by the International Union for Conservation of Nature affirming that every form of life warrants respect independently of its worth to people and that human development must not threaten the integrity of nature or the survival of other species.

Boyd writes that in a dramatic development spanning the globe, non-human animals and ecosystems are being recognized as legitimate rights-bearing subjects. "The fundamental values and laws that have governed society for hundreds of years are in the early stage of the most radical transformation in history. . . . Recognizing that animals and other species have rights rejects anthropocentrism, challenging the global predominance of a single species. Recognizing that nature itself has rights goes even further, undermining the idea of property, and bringing into question our wholesale and accelerating appropriation of the planet."[335]

In an April 2025 report, *Inside Climate News* explained that evolutionary biology shows that humans share a common ancestor with all other life on Earth that must be protected in law:

> "Rights of nature" is a movement aimed at advancing the understanding that ecosystems, wildlife and the Earth are living beings with inherent rights to exist, evolve and regenerate. Legal rights are the highest form of protection in most governance systems. In the United States, humans and non-humans have enforceable legal rights, like corporations' right to freedom of speech. At the same time, most legal systems treat nature as rightless property that humans can own, use and destroy. That means the law views sentient species like elephants and bald eagles, as well as life-supporting ecosystems like forests and coral reefs, no differently than objects like microwaves or cars.

Environmental law is a relatively young field, explained the nonprofit news agency and is flawed in that it is designed to permit pollution. Laws only control how much. "Rights of nature laws start from an entirely different place. Ecosystems, wildlife and Earth itself are treated as living beings with inherent rights deserving of the highest form of legal protection. The central concern of rights of nature laws is to maintain and preserve the integrity of ecosystems, requiring governments to take a preventative, rather than a reactionary, approach." Rights of nature are enforced through a longstanding concept called legal personhood, with human "guardians" representing nature in court, a similar arrangement used for minors in court proceedings.[336]

In 2010, the Global Alliance for the Rights of Nature (GARN) was founded, bringing environmental, Indigenous, human rights and conservation organizations together with scientists, economists, writers and spiritual leaders looking to transform our human relationship with our planet. GARN acknowledges that nature in all its life forms has the right to exist, persist, maintain and regenerate. The network collects and reports on the hundreds of examples around the world where the rights of nature are being recognized. Some, such as Bolivia and Ecuador, have brought rights of nature into their constitution. Many others are legal findings of local or national courts. Others are community-based, initiated by local groups — often Indigenous — and adopted by local governments.

The Global Alliance for the Rights of Nature is clear that, to ensure an environmentally sustainable future, humans must reorient themselves from an exploitative and self-destructive relationship with nature to one that honours the deep interrelation of all life and contributes to the health and integrity of the natural environment. For this to happen, we need to establish a system of jurisprudence that sees and treats nature as a fundamental rights-bearing entity and not as mere property to be exploited at will. This "needed system of jurisprudence" is already underway, guided by local communities, municipal councils, Indigenous groups and, in some cases, state, provincial and national governments as well.

There are now close to 500 rights of nature projects in 44 countries, with many new initiatives on the way.[337] UN Secretary-General António Guterres says that Earth jurisprudence has emerged as the fastest-growing legal movement of the 21st century. Ecuador recognized the rights of nature in its constitution in 2008 and has since ruled in favour of mangroves, cloud forests (in a case that stopped mining in the Los Cedros reserve) endangered frogs (in a case that stopped the construction of a copper mine) and coastal marine ecosystems. In a July 2024 decision, an Ecuadorian court ruled that pollution has violated the rights of the river running through the country's capital, Quito, and ordered the government to come up with a cleanup plan.

New Zealand is another pioneer in the rights of nature movement. In 2014, it granted personhood rights to Te Urewera, a vast native forest on the North Island. In 2017, it granted legal rights to the 290-kilometre Whanganui River and assigned a Māori community as the river's guardian. And in January 2025, Mount Taranaki, the North Island's second highest mountain and considered an ancestor by the local Indigenous people, was recognized as a legal person and granted all the rights and responsibilities of a human being. This was a deeply important decision for the local Māori, as, in the past, they had been banished from much of the mountain and surrounding land to make way for hunting and tourism. "Today, Taranaki is released from the shackles of injustice, of ignorance, of hate," said Debbie Ngarewa-Packer, descendant of the first Taranaki tribes.[338]

Lake Titicaca is the highest navigable body of water in the world, situated in the Andean mountains on the border of Peru and Bolivia. Despite being a UNESCO World Heritage Site and a sacred waterway to the local people, the lake is dangerously contaminated. Women from the Aymara and Quechua communities took initiative to protect and restore the lake, and in April 2025, the regional government of Puno, Peru, officially recognized Lake Titicaca as a "subject of rights," including the rights to exist, thrive, be restored and regenerated.

Rivers are being granted rights all over the world. Hundreds of organizations have signed a petition to urge governments at all levels to adopt the Universal Declaration of River Rights. It states that all rivers are living entities, entitled to fundamental rights as well as legal guardians and whose rights extend to the health of all watersheds and river basins.

Citizens of Everett, Washington, voted to enshrine its watershed's right to exist, regenerate and flourish into municipal law in 2024. The same year, a Peruvian court named the Kukama people the guardians of the now legally protected Marañón River in northern Peru. The declaration of rights of the Biobío River in Chile's Patagonia region is the first river protection charter in that country. In West and East Sussex, England, a local council has voted to enshrine the Rights of River Charter to protect the River Ouse "for generations to come." A petition is being widely circulated in Scotland to grant the River Clyde its legal right to personhood. In November 2024, Spain's Constitutional Court upheld a law giving legal personhood to the Mar Menor, Europe's largest saltwater lagoon that sits on the country's Mediterranean coast. There is currently a referendum drive in Lucerne, Switzerland, to recognize the legal rights of the River Reuss. And in the summer of 2025, Paris announced it is seeking legal personhood for the River Seine in response to the recommendations of a citizens' convention on the future of this famous river. In March 2025, a bill was tabled in the New York State Assembly that would guarantee the legal rights of all water bodies in the state against destruction and recognize "unalienable and fundamental rights to exist, persist, flourish, naturally evolve, regenerate and be restored" for the Great Lakes and other watersheds and ecosystems throughout New York State.[339]

In Canada, in addition to the Magpie River in Quebec, rights of nature initiatives underway include Lake Winnipeg in Manitoba;

the Gatineau River in Western Quebec; the Athabasca River Basin that spans 159,000 square kilometres in northern Alberta; the Yukon River, whose drainage area is almost 855,000 square kilometres; and the St. Lawrence River shared between Canada and the United States that connects the Great Lakes to the North Atlantic Ocean. In October 2024, the British Columbia Assembly of First Nations adopted a resolution calling on the province and others to work with First Nations to advance the legal personhood of nature, "including water bodies such as rivers and lakes."

In most cases, the Canadian projects are led by Indigenous groups. Nika Silverfox-Young, a citizen of the Little Salmon Carmacks First Nation in the Yukon, told the CBC her peoples' connection to the river is "blood memory" and the river is "part" of her. "We really travelled a lot and we used the rivers as our highways. We really needed them. We'd pick out the best parts and listen to the land when the salmon were coming back. We've lost that ability to listen to what our land is telling us. She is screaming right now about how we need to protect her. . . . I feel like I'm legally, genetically obligated to protect this land."[340]

In his 2025 book, *Is a River Alive?*, British author and academic Robert Macfarlane writes that we have forgotten that our fate flows with that of rivers and always has. "Our relationship with fresh water has become intensely instrumentalised, privatised and monetised: river understood as resource, not life force. The duty of care for rivers, who extend such care to us, has been abrogated. Regulation has gone unenforced, monitoring is strategically underfunded. . . . We now take it for granted that we take rivers for granted. It is unremarkable that a company registered yesterday is, in the eyes of the law, an entity with legal standing and a suite of rights, including the right to sue — but that a river who has flowed for 10,000 years has no rights at all." It has long been in the interests of power to deem nature dead, he writes, in preparation for its extraction, conversion and consumption. But just as rivers are easily wounded, given a chance, they heal themselves with remarkable speed: "Their life pours back."[341]

And the movement is expanding into new frontiers. The Earth Law Center, a US-based team of Earth lawyers from around the world, has pioneered Ocean Rights, a legal framework and ethical perspective toward recognizing the oceans' inherent rights and intrinsic value. Along with Ocean Vision Legal and others, the centre is calling for a Universal Declaration of Ocean Rights to change the status quo in ocean governance. The Brazilian city of Linhares has legally recognized waves as living beings, making it the first time that part of the ocean has been granted legal personhood. In early August 2024, it passed a new law that gives the waves at the mouth of the Doce River, which runs to Brazil's Atlantic coast, the intrinsic right to existence, regeneration and restoration.

In 2024, *Inside Climate News* reported on the initiative to recognize Antarctica as an autonomous legal entity with rights to participate in climate discussions that impact its future and the planet. Advocates argue the continent's unique role in Earth's stability justifies legal protection, allowing representation at events like UN climate talks.[342] And in an exciting development, on Mother's Day 2025, the youth hub of the Global Alliance for the Rights of Nature called on the United Nations to adopt a Declaration of the Rights of Mother Earth and signalled their intention to promote it among youth.

Of course, many of these exciting projects are taking place alongside continued pollution and destruction of land and water, even in countries whose constitutions adopted a rights of nature amendment. Rivers that are "protected" are still facing harm. It will take time to turn the tide, but the grassroots nature of this movement gives great hope and a counternarrative that we are not helpless against the forces of destruction.

Promote a rights-based approach to conservation

An exciting new development is an emerging consensus that two formerly separated areas of law — biodiversity law and human rights law — must come together to protect both. Simply put, these rights are intertwined. Humans cannot live on barren land that can no longer sustain them, no matter what human rights laws exist to supposedly protect them. Healthy forests, soil and biodiversity support the enjoyment of a wide range of human rights. Conversely, protecting the rights of the people who understand their ecosystem and how to care for it helps the natural world to survive and thrive. When whole communities live without the human rights to sanitation and municipal services, they are forced to use local streams and rivers for human and industrial waste, thereby destroying the rights of those waterways.

As a January 2025 report by the German Institute for Human Rights explains, "The loss of biodiversity threatens the enjoyment of human rights because the destruction of resources and ecosystem services has significant impacts on peoples and communities. . . . Failure to protect human rights can trigger environmental degradation, because ignoring the needs and voices of individuals and communities who can contribute to conservation will result in a less protected environment."[343]

Forest Peoples Programme, a human rights organization with consultative status at the UN, works with forest people across the globe. In a 2022 report to the UN, the organization defines their approach: "A human rights-based approach means, in simple terms, that biodiversity policies, governance and management do not violate human rights and that those implementing such policies actively seek ways to support and promote human rights in their design and implementation. Indigenous peoples and local communities' ways of life and territories are part of the solution to our global crises and must be identified and supported, including through recognition of rights over lands, territories and resources, in area-based measures, in customary sustainable use, in traditional knowledge and in full and effective participation."[344]

As David Boyd, then UN special rapporteur on human rights and the environment, wrote in a 2021 policy brief, "fortress conservation" is motivated by the mistaken belief that successful conservation outcomes require "pristine wilderness" free from human inhabitants.

He argued that a human rights approach to conservation protects not only the affected human communities but ecosystems as well and is the only way forward:

> In addition to being morally and legally required, human rights–based conservation is the most effective, efficient, and equitable path forward to safeguarding the planet. Mounting evidence confirms that Indigenous Peoples and other rural rights holders possess the knowledge and ability necessary to successfully conserve and manage biodiverse ecosystems more effectively than governments and at a fraction of the cost, particularly where their rights (including the specific rights of Indigenous and rural women) are recognized, respected and supported. Indigenous Peoples and other rural rights holders steward and claim collective rights to over half the global land area, employing customary tenure systems anchored in both traditional and contemporary knowledge to successfully manage and conserve vast ecosystems. When these communities' impressive conservation capacities are considered alongside the extensive contributions of other marginalized rural rights holders, such as smallholder peasant farmers whose agro-ecological knowledge and practices provide a viable and inspiring alternative to the industrial food system responsible for the bulk of global biodiversity loss, the potential power of rights-based solutions to combat the global biodiversity crisis is clear.[345]

Helen Tugendhat, the co-executive director of Synchronicity Earth, told *Mongabay* that shifting the dynamics of power by giving legal land titles and management authority to people living in protected landscapes will produce better ecological results. "What the previous approach has always missed is, why try to either move or constrain the people who are already there and bring in other people who then have to manage the area in a different way? That's an incredibly expensive process, and when there isn't enough money to balance national budgets, protected area budgets aren't the ones that get prioritized. We should be investing in the people already customarily managing those areas instead." She also called for increased support and protection of environmental defenders as well as strict conditions on climate and biodiversity funding.[346]

David Boyd writes that one of the most interesting developments comes from Latin America, where governments, communities and the courts are beginning to merge the right to a healthy environment with the rights of nature. The Inter-American Court of Human Rights as well as the higher courts of Mexico, Colombia, Ecuador and Panama are making it clear that the right to a healthy environment has two aspects — it must be healthy for humans *and* nature. "In our quest to protect and recover the biodiversity and ecosystems that are the life support system of planet Earth and one of the great wonders of the universe, the only effective and equitable path forward is putting human rights and nature's rights at the very heart of every conservation action that is taken," he explains.

There was real progress at the Biodiversity COP15 held in Montreal in 2022, where human rights were enshrined in the framework document. The Parties called Indigenous Peoples and local communities "custodians of biodiversity" and "partners in conservation, restoration and sustainable use" and called for "their full and effective participation in decision-making" and "free, prior and informed consent" in the implementation of biodiversity protection and restoration, in applying the 30 x 30 commitment. They also recognized the "rights of nature and rights of Mother Earth" as being integral to successful implementation of the pledges. The framework also acknowledges the human right to a clean, healthy and sustainable environment.[347]

While this was indeed a breakthrough, there were still deep divisions over the issue of biodiversity offsets at the November 2024 COP16, held in Colombia. Nevertheless, the voices of those in opposition were clearly prominent and heard. Friends of the Earth International and the Global Forest Coalition rejected this false solution and called for rights-based solutions that prioritize the protection of biodiversity and local communities. It is clear from the official statements coming out of these international gatherings that the concerns of the critics of the financialization of nature are being taken seriously.

The German Institute for Human Rights is one of many calling for governments and international institutions to use a rights-based approach to protecting biodiversity. "Moving forward, a robust commitment from States to implement and enforce these standards is essential. Strengthening biodiversity protection through human rights law not only supports planetary sustainability but also empowers those most affected by environmental degradation, fostering a more resilient and equitable approach to preserving our planet's biological diversity."[348]

Indeed, in a welcome move, the United Nations Environment Programme (UNEP) has released a report outlining existing human rights standards and how they should be applied to private conservation organizations and funders in their conservation operations. Recognizing the biodiversity crisis that is upon us, UNEP acknowledges that vulnerable people and communities around the world often suffer from the actions of conservation organizations and funders.

"To be successful, efforts to conserve nature must recognize the rights and contributions of those whose ways of life contribute to its protection. In particular, the rights and methods of managing and conserving nature of Indigenous Peoples, Afro-descendant and other tribal communities, and peasants should be prioritized," it asserts. Core principles include respecting human rights, engaging in good faith consultation with local communities, providing for effective remedies where there is damage to local people and protecting against human rights violations in law enforcement.[349]

Protect the water

We urgently, desperately need to protect and restore freshwater everywhere on the planet. We need to set strong rules around access to groundwater and get deadly serious about stopping the pollution of our lakes and rivers. We need to restore lost urban rivers, create "sponge cities" to capture rainwater and storm water, protect wetlands and forests — the kidneys and lungs of our watersheds — and bring our watersheds back to health. We need to do this not only because the planet is quickly running out of accessible clean water, putting the lives of untold millions at risk, but because healthy watersheds and water cycles cool the planet and help us deal with the ravages of the climate crisis.

A powerful movement has grown up around the world to harvest and retain rainwater to restore water-related ecosystems destroyed by poor farming practices, deforestation and wetland removal. Simply put, when vegetation or water is removed from an ecosystem and there is no water to evaporate or transpire from the landscape, it rains less and heats up more quickly. Aquatic scientists and experts explain that a key antidote to the climate crisis is to bring back water and rain to watersheds and deserts alike, reestablishing the atmospheric moisture cycle necessary to restore precipitation patterns and cool the planet.

Examples abound. Hydrologist Michal Kravčík and others have led an initiative in Slovakia that the UN says brought back water and revived soils. Using a system of small dams and weirs, bioretention ponds, infiltration trenches and buffer strips, the team recently reclaimed 27 hectares of arid land and created a Bioclimate Park, a vibrant living farm and an outdoor climate discovery trail nestled in the Carpathian Mountains' Rajec Valley. Over the last 20 years, Kravčík and his team have set up thousands of water-retention schemes in hundreds of Slovak communities.

Water harvesters say we are losing vast amounts of freshwater to the oceans every day when rain falls on urban pavement and soil hardened by chemicals and overuse. Kravčík reports that the temperature varies in the cities of his country by as much as 20 degrees, depending on whether they have adopted water-retention practices. He predicts that the average global temperature would drop by more than 0.5 degrees if all the cities of the planet took part. Salisbury, South Australia, has greened the desert by creating dozens of wetlands that capture rainwater and purify stormwater with carefully selected species of plants. The water is then stored in aquifers for industrial and irrigation use. Wildlife and birds have returned to this green oasis.

The Great Green Wall initiative was launched in 2007 to create a wall of greenery spanning the width of Africa to keep the Sahara Desert from encroaching on the rest of the continent. As reported by *Upworthy*, one project in Senegal has had incredible success with the rejuvenation of 300,000 hectares of what was thought to be unusable land by using long-forgotten techniques for harvesting water. Just as their ancestors had done, communities built half-moon-shaped ditches to capture water and transformed their desert into thriving farmland.[350]

At the March 2023 UN water conference in New York City, several African and Latin American countries launched the largest river and wetland restoration project in history. Led by a coalition of governments that include Colombia, Mexico and Gabon, the plan aims to restore 300,000 kilometres of rivers — as long as seven times the Earth's circumference — and an area of wetlands larger than India by 2030, reported *Agence France-Presse*.[351]

Water restoration can happen in cities as well. According to *The Guardian*, in the 1960s, Switzerland had among the dirtiest water in Europe, "blighted by mats of algae, mountains of foam, scum, and dead fish floating on the surface." Swimming was banned as people got ill if they swallowed river water. Through strict laws and action to ensure wastewater treatment, water purification — including the removal of micropollutants — and water protection, the rivers and lakes in Swiss cities are so clean the citizens swim and fish in them every day.[352]

"Daylighting" is the opening up and restoration of rivers and streams that have been paved over in cities. Freeing these buried waterways returns the natural environment of the river and reduces runoff, creates habitat for species and creates urban oases. Zurich is a pioneer and has daylighted over 21 kilometres of buried streams to date. In Seoul, the 11-kilometre Cheonggyecheon River, once covered by a highway, was restored in 2005, creating a beautiful river park in the heart of the city and bringing back fish and bird species not seen in years. The writer Toni Morrison said, "All water has a perfect memory and is forever trying to get back to where it was."

Montreal, Canada, is a "sponge city," a project initiated in China that manages stormwater and reduces flooding by creating natural infrastructure to absorb and retain rainwater. The city is building dozens of sponge parks to soak up excess rain to help it deal with the flooding that has increased due to climate change. "Instead of seeing the rain as something bad, we are now keeping the rain and turning it into something beautiful," Philippe Sabourin, a spokesperson for the city, told the CBC.[353] The sponge city movement is growing, with at least 70 in China alone.

Another positive development is the movement to remove dams to enhance river health, build climate resilience and benefit people and nature. The World Wildlife Fund reports that flowing rivers are critical to global efforts to tackle the nature and climate crises as they help navigate ever-increasing droughts, floods and storms while boosting biodiversity. But across the world, rivers are choked with dams, weirs and culverts, fragmenting them and disrupting the natural flow of water, sediments, nutrients and species, undermining their resilience, impacting subsistence fisheries and accelerating nature loss.

> Dam removals are a fast, proven, and cost-effective solution with lasting benefits for communities and nature. Over ten thousand barriers — from giant hydropower dams to weirs and culverts — have been removed already, primarily in the US and Europe. By reconnecting rivers, dam removals have helped to build resilience and bring life back to dying rivers — from Finland to France, UK to Ukraine, and Maine to Spain. Dams are also starting to be removed in parts of Asia, Oceania and South Africa. And a movement is growing with increasing numbers of people realizing that dams are not forever — and that restoring rivers is critical to our future.[354]

In November 2023, European member states reached an agreement to remove man-made barriers from 25,000 kilometres of free-flowing rivers by 2030.

In summer 2024, the largest dam removal in US history was completed with the removal of the last of four dams on the Klamath River, reopening more than 644 kilometres of river in southern Oregon and northern California. This removal restored 2,200 acres of land that is above ground for the first time in a century. The local Indigenous community, the Yurok, harvested and nurtured billions of native seeds that they then planted on the newly released land. The Yuroks had been displaced from their land to make way for the dams and are very proud that they led the fight to have the dams removed. "Taking care of a plant that your great-grandmother took care of puts a lot of perspective into life. And this is the first step in reclaiming our ability to take care of the Klamath River," tribe member Brook Thompson told the BBC.[355]

All these initiatives require democratic governance of water and water systems, something that cannot be assured if water is privately owned or controlled. Just as people and communities have the right to proactively protect and restore their waterways, so too do they have the right to demand their governments establish tough environmental rules. We cannot protect a watershed if a giant factory farm is allowed to continue to pollute it by offsetting its effluent with "nutrient trading" credits. We cannot protect a river if it has been sold to an asset manager or a lake if a mining company has 50-year rights to dump its tailings into it. And we cannot promote a rights-based approach if a for-profit transnational water utility is running urban water services.

The enclosure of the water commons is destroying communities and denying people the right to water and sanitation. This right cannot be fulfilled under a market or corporate model of water governance. To ensure that the world's dwindling water sources are protected and more equitably shared, we must promote conservation, water justice and democracy. If we continue to allow commodification of the world's water, it will flow toward those with money and power, not to the communities and ecosystems that need water for survival.

We must be clear that water is a commons, a public trust and a human right. As well, water has rights in and of itself outside of its service to humans. To protect both the human and ecological rights of water, we must adopt a system of public trust, whereby all competing uses for a watershed or aquifer have to pass a test, not just of fairness of access today, but also the future capacity of the water body. Public trust offers principles that combine public good, control and oversight with long-term protection of the watershed. It also sets the stage for an agreed-upon "hierarchy of use" whereby some uses of the water — such as for essential human needs and ecosystem protection — will take preference over others.

There is a rich history of public trust in US law. In 2008, concerned about major groundwater extractions, Vermont passed a groundwater protection act that declared its groundwater to be a public trust resource, legally belonging to all Vermonters, that must be managed in the best interests of all. A permit system has been set up for those who use over a certain limit per day, and the state has the right to revoke these permits if they are abused. Several years after adopting it, the Vermont Natural Resources Council used the state's public trust legislation to challenge a tritium leak from the Vermont Yankee Nuclear Power Station, declaring that violation of the integrity of the water was a violation of the rights of the owners — the people of Vermont.

Pedro Arrojo-Agudo, the current UN special rapporteur on the rights to safe water and sanitation, said in an interview with a Spanish cooperative, "There are two big challenges: the first is making peace with our rivers and our aquatic ecosystems if we really want to make progress so that these millions of people can have clean water. The second major challenge is the democratic governance of water. To understand water not as a commodity, but as a common good, accessible to all, not to be appropriated by anyone. When we understand it in this way, the priority and the human rights approach become clear."

Arrojo explained if we want to protect the human right to water, we need a "water transition" toward a new water culture that requires democratic governance and sustainability of ecosystems:

> The priority is to sustain life, biodiversity and human dignity. . . . I argue that the water available in each territory should be considered a common good, and that the aquatic ecosystems that supply us should be considered a common natural heritage. What does this mean? It means that, above all, we must give priority to the principles of sustainability and equity. The market is not useful for this. . . . And that is why the participation of the communities involved, who depend on this river, on the water available in their environment, is fundamental. It should not be managed by the market, because then the poor do not have human rights. It is essential to articulate governance

> that is participatory, transparent, accountable and open to broad and effective citizen participation.[356]

Many positive steps have been taken to maintain democratic control of water and protect the rights of communities to be its guardians. In 2010, the UN General Assembly and Human Rights Council recognized the human rights to water and sanitation, a critically important advance that has helped people around the world to protect local water sources and services. Dozens of countries have either amended their constitutions or introduced new laws to enshrine the right to water. As well, the fight against privatized water services has been very successful in Europe and North America. More than 300 cities that privatized their water services realized their mistake and brought these services back under public management.

Some of these cities, including Paris and Berlin, have also become Blue Communities, taking the pledge to promote water as a human right, maintain water as a public trust by keeping it under public governance, ban the provision of bottled water where clean public water is available and promote public water services in the Global South. Blue Communities originated in Canada in 2009, when the government of the day was promoting the privatization of municipal water services, but the Blue Communities movement soon spread to Europe and other parts of the world. There are close to 100 towns and cities around the world that have taken the pledge and many universities, unions, faith-based communities and schools have joined the movement.

Roland Brunner, a retired Zurich public sector union organizer, coordinates the Blue University community, keeps records on all the Blue Community projects around the world and publishes a Blue Community newsletter. In an email exchange, he said the movement gives him hope. "It is not necessary to explain to people why water and sanitation are key to human well-being. But there is a need to give people a way to advocate for these essential assets, to counter their sense of hopelessness and frustration in the face of global climate change and environmental disasters, to take action and live hope rather than wait for policies and politicians. The Blue Community offers a simple, informal but highly valuable way to get involved and participate in a global movement and network. To stop the destruction of this blue planet, everyone can take action."

We Own It, a British civil society organization that fights the privatization of water services, reported that 82% of the people of England support public ownership of water and believe that the country is ready, after decades of abuse at the hands of private water utilities, to take its water services back under democratic control.[357] This would be a major victory considering that country was one of the first to privatize its water, and cities like London would be warmly welcomed into the Blue Communities family.

The People's Water Forum is a network of water justice movements around the world. Marcela Olivera is a Bolivian water activist and coordinator of the Blue Planet Project, an international organization working for the human right to water and a founding member of the People's Water Forum. At the 2023 UN water conference in New York City, Olivera presented the Water Justice Manifesto on behalf of the forum to the assembled participants. The manifesto beautifully offers a clear definition of the dual rights — human and ecological — that must guide us as we move forward to protect water for people and the planet:

> Access to water and sanitation are fundamental human rights. Water is a common good, and must be accessible to all without discrimination, under public control and not a commodity. Personal and domestic uses of water, including for hygiene, should have the highest priority over productive uses, such as agri-business and industry.
>
> Water policies must prioritize the sustainable management of rivers, lakes, wetlands, springs, and

> aquifers, guaranteeing their good ecological status, within the framework of the human right to a healthy environment and as key to confronting ongoing crises of pollution, deforestation, desertification, biodiversity loss, and climate change. Governments must ensure that agri-business and industrial users are accountable and responsible for their use of and impact on all natural resources, including water, based on legislation, regulation and enforcement and not rely on voluntary measures.[358]

Promote strong international law to protect climate and biodiversity

The movers behind the Paris climate agreement were not wrong. The brilliant and dedicated experts and scientists of the Intergovernmental Panel on Climate Change are not wrong. António Guterres, UN secretary-general, is not wrong. The world is in the grips of a climate crisis, a crisis of our own making that threatens our very survival. "In the case of climate, we are not the dinosaurs, we are the meteor," Guterres told a June 2024 gathering at the American Museum of Natural History. "We are not only *in* danger, we *are* the danger."[359]

Past commitments to deal with the climate crisis have been put on hold or overturned as climate has been overtaken by other political and economic agendas and a small but vocal and influential minority continue to deny that a climate crisis exists. Only 15 countries met the February 2025 deadline to submit their Paris Agreement plans to slash greenhouse gas emissions — just 8% of the countries who signed the pledge. Some of the largest emitters failed to submit reports, including China, India and Russia. This led University of Westminster geographer Doug Specht to speculate on whether the era of cooperation on climate change is over:

> The failure of most nations to submit new emission plans suggests that the era of cooperation on

> climate change is over. The largest and most powerful of these nations are growing their military and diplomatic presence around the world, particularly in countries with large reserves of critical minerals for electric vehicles and other technology relevant to decarbonization. The lack of [Nationally Determined Contributions] from these nations may be less a matter of middling green ambitions, more an attempt to disguise their planned exploitation of other countries' resources.[360]

And then, of course, there is President Trump, who pulled the US out of the agreement altogether. Soon after he took office, he launched an all-out assault on climate protections. He championed coal, killed renewable energy projects, gutted the Environmental Protection Agency and cut critical research funding for climate action. In April 2025, Trump ordered NASA to end funding for the US Global Change Research Program that produced the federal government's preeminent climate report. "It is not an overstatement to say that the Trump administration has launched the worst White House assault in history on the environment and public health. Day by day and hour by hour, the administration is destroying one of the signature achievements of our time," said Manish Bapna of the Natural Resources Defense Council. "If this assault succeeds, it could take a generation or more to repair the damage."[361]

As disturbing as this is, we must not give in to discouragement, but instead, renew our commitment to ending our dependence on fossil fuels and creating a clean energy future. Guterres says we have an "exit ramp off this highway to climate hell," as we still have control of the wheel. He called on every country in the world to institute a ban on advertising from fossil fuel companies and to invest in clean energy. He also calls for "climate justice," recognizing the most vulnerable nations and their people have been left stranded with a crisis they did little to create. "We cannot accept a future where the rich are protected in air-conditioned bubbles while the rest of humanity is lashed by lethal weather in unliveable lands."[362]

The Fossil Fuel Non-Proliferation Treaty Initiative, a project of Earth Island Institute, is a new effort to foster international cooperation and accelerate a transition to renewable energy. Organizers say the initiative is backed by a growing network of governments, civil society organizations, academics, scientists, youth activists, Indigenous Peoples and faith-based communities. In May 2024, the initiative held its first meeting in Antigua and Barbuda and is attracting the support of hundreds of cities, thousands of organizations worldwide, a growing number of nations and millions of committed people.

Canadian energy policy analyst and activist Tzeporah Berman is a founder of the initiative and says it is time to name the culprits — the fossil fuel companies and the political leaders who support and subsidize them. Berman directly links emissions from the world's largest fossil fuel companies to the increasingly intense forest fires in the western US and Canada. She lauds the State of California and the city of Los Angeles for joining the call for a fossil fuel non-proliferation treaty. She says such a treaty would foster multilateral cooperation around a binding agreement to end the expansion of fossil fuels, wind down existing production to levels that keep climate change in check and work together to ensure a just transition:

> My hope for the future is that we're able to see across race, class and political differences and come together with a common sense of purpose. We need to build a movement that includes millions who are standing up together to call for the future that we want, a cleaner and safer future. . . . In some ways, I feel like the current climate debate has robbed us of our imagination. We do all these models and assessments. But we fail to imagine a scale of change and

> leadership that is totally possible. If we're going to create that future, we have to imagine it.[363]

Similar movements are coming together to spur international action on the plastics crisis. At the end of the UN Environment Assembly in March 2022, representatives from 175 nations agreed on a mandate to create a legally binding treaty to end plastic pollution. As addressed in previous chapters, they have so far failed to agree on the terms, as many countries want to cap production while a handful of producing countries, prodded by industry lobbyists, are only prepared to target plastic waste. Meanwhile, plastic's toxic footprint continues to grow unabated.

However, there are several civil society movements that show great promise. Break Free from Plastic brings together more than 13,000 organizations and individuals around the world to demand reductions in single-use plastics and advocate for lasting solutions to the plastic pollution crisis. Break Free fights petrochemicals, pushes for corporate accountability, promotes systemic solutions and builds zero-waste communities. It is clear in its opposition to a market solution. "As the world's governments come together to negotiate a new international treaty to tackle plastic pollution across its lifecycle, it is vital that the reality of plastic offsetting is understood. A concept shrouded in smoke and mirrors should not be incentivised in a treaty designed to reduce plastic pollution," it says in its 2023 report, *Smoke and Mirrors*, on plastics offsetting and credits.[364]

Greenpeace has initiated an international campaign promoting a global plastics treaty that would limit both production and use. It is organizing "an unstoppable global movement" of millions to end the age of plastic. Greenpeace is seeking a treaty that cuts plastic production by at least 75% by 2040 and ends single-use plastic. It says that the treaty must be built upon a foundation of human rights and seek to reduce inequalities, end waste colonialism, prioritize human health and ensure dignity for all. The treaty must also protect biodiversity, safeguard our climate and ensure a just transition to a reuse-based economy.

Greenpeace has been particularly involved in the African campaign to ban single-use water sachets. These small heat-sealed, water-filled, polyethylene plastic bags are ubiquitous and are burying the streets and clogging the water and drainage systems of cities while poisoning communities across the continent. Water sachets are popular as the public water in many places is unfit to drink. The UN Environment Programme estimates that about 50 to 60 million used water sachets are thrown into the streets daily in Nigeria alone. Greenpeace is working with youth activists to ban water sachets, and 34 African countries have pledged to do so.[365]

Peoples' movements have been influential in the campaign that led to the adoption of the Global Oceans Treaty by the United Nations in June 2023. The treaty, also referred to as the "High Seas Treaty," allows governments to create ocean sanctuaries free from destructive human practices and promotes the creation of vast ocean sanctuaries beyond territorial boundaries where marine life can recover and thrive. The goal is to protect 60% of the Earth's non-territorial oceans from harmful practices, including pollution, unsustainable fishing practices and warming temperatures. One hundred and forty-five countries signed the treaty, and as of September 2025, 69 had pledged ratification, ensuring the treaty will become legally binding in 2026.[366]

Greenpeace is also involved in this campaign and was the first to call for a treaty in 2005. In 2023, it published a major report called *30 x 30: From Global Ocean Treaty to Protection at Sea*, setting out the political process to deliver protection of the Earth's oceans, and recommending three specific sites on the high seas to be the first set of ocean sanctuaries due to their ecological significance: the Sargasso Sea, the Emperor Seamounts in the Northwest Pacific Ocean and the South Tasman Sea/Lord Howe Rise between Australia and New Zealand. At the 2025 UN Ocean Conference, French Polynesian President Moetai Brotherson announced the world's largest marine protected area, spanning 4.8 million square kilometres of ocean around its 100-plus islands in the South Pacific.

In Canada, more than 30 Canadian conservation groups, including Pacific Wild, Living Oceans, Nature Canada, David Suzuki Foundation and West Coast Environmental Law, are calling on the Canadian government to show leadership by committing to sign and ratify the treaty. They wrote in a November 2023 letter to the Canadian government:

> The future of Canada's oceans relies on more than what occurs within our jurisdictional boundaries alone. Because of this, we as a country must be willing to work with countries around the world on the common goal of protecting and conserving ocean ecosystems. . . . By taking these steps, the Government of Canada will show the rest of the world their support for ocean protection and desire to contribute as ambitious allies to global ocean conservation goals, including the goal to protect 30% of the world's land and ocean by 2030, the goal to halt and reverse biodiversity loss, the commitment to uphold the UN Sustainable Development Goals, and Canada's recent support for a moratorium on deep sea mining in international waters.[367]

Astrid Puentes Riaño, UN special rapporteur on the human right to a healthy environment, links the rights of oceans to the rights of marginalized humans and communities that depend on them, thereby promoting a rights-based approach to ocean restoration. In a March 2025 report to the Human Rights Council, she said that ocean degradation threatens communities and affects human rights worldwide, including the right to a healthy environment. "Knowing the interdependence and interconnectedness of humans and ecosystems with the ocean is essential to understanding the current impacts on this delicate balance, even for those living inland," she said of the report.

"The ocean is the largest biome on Earth, covering 70 per cent of its surface. One third of the human population (2.4 billion people) live within 100 km of an ocean coast. Despite over 600 agreements, marine ecosystems face pressing threats including climate change, overfishing, extractivism, pollution, and deep-sea mining." These issues are exacerbated by "weak governance and enforcement gaps; disproportionate impacts in Indigenous Peoples, small-scale fishers, and coastal communities; escalating violence against ocean defenders, and insufficient accountability." She called for strengthening legal protections for marine biodiversity and coastal communities; implementing stricter regulations on overfishing, pollution and offshore extractive industries; and enforcing the precautionary principle, all while recognizing the role of ocean defenders and Indigenous knowledge in marine governance.[368]

The right to a healthy environment was adopted by the UN General Assembly in 2022, bolstering international jurisprudence that connects the right of people to the health and rights of the planet. While the resolution is not binding, 161 countries voted in favour. More than 100 countries have added the right to a healthy environment to their constitutions and many have recognized it through new laws. Canada added this right in 2023 when it updated its signature environmental law, the Canadian Environmental Protection Act. In July 2025, the International Court of Justice issued a unanimous advisory opinion, recognizing the human right to a clean, healthy and sustainable environment and held that states have a legal obligation under international law to protect the climate system and prevent environmental harm.

Once again, the impetus behind this important UN milestone was the hard work of environmental and human rights organizations, including Indigenous Environmental Network, Earthjustice, Greenpeace, Natural Resources Defense Council and Friends of the Earth International. Canada's David Boyd, who was serving as

UN rapporteur at the time, spearheaded the campaign at the UN. "These resolutions may seem abstract, but they are a catalyst for action, and they empower ordinary people to hold their governments accountable in a way that is very powerful," he said.[369]

In 2014, the UN Human Rights Council started developing a legally binding instrument to hold transnational corporations accountable for human rights violations. Again, civil society, especially the trade union movement, was instrumental in getting the project moving. The Washington- and Switzerland-based Center for International Environmental Law reported in 2018 that more than 600 organizations from around the world were calling for this treaty.[370]

The World Bank announced a positive initiative in May 2025, when it released its first-ever framework to address environmental and social harms caused by projects the bank has financed through its private sector branches, such as the International Finance Corporation and the Multilateral Investment Guarantee Agency. Some were disappointed that the new rules will only apply to future projects and not be available for past grievances. Nonetheless, the move was greeted as a sign that this institution is finally listening to its critics. "This is historic. It's the first actual directive mandate for the IFC that says when one of the projects they finance causes harm, they have to provide remedy," Carla García Zendejas, director of the people, land and resources program at the Center for International Environmental Law, told *Mongabay*.[371]

A recent proposal to put environmental destruction on the same level as genocide, crimes against humanity and war crimes, is causing a huge stir. The South Pacific country of Vanuatu, backed by Fiji and Samoa, is pushing to have ecocide recognized by the International Criminal Court in the Hague. The CBC reports that the term "ecocide" was first used in 1970 in reference to the damage done by the US military's use of Agent Orange during the Vietnam War. Had it been in place, the charge of ecocide could have been extended to incidents like Russia destroying the Kakhovka Dam in

Ukraine in 2023, or former Iraqi president Saddam Hussein ordering the spilling of 11 million barrels of crude oil in the Persian Gulf in 1991. The CBC quotes Jojo Mehta, co-founder and CEO of Stop Ecocide International, who said, "If we put ecocide alongside crimes against humanity and genocide, what we're saying is it's just as bad, wrong, dangerous to severely destroy ecosystems as it is to commit those other crimes."[372]

And of course, there is the intense struggle referred to in earlier chapters about the future of the Convention on Biological Diversity, adopted at the Rio Earth Summit in 1992 and the Kunming-Montreal Global Biodiversity Framework adopted in 2022. As mentioned earlier, the growing realization of the need to address biodiversity and ecosystem health and restoration in confronting the climate crisis is both welcome and urgent. This process is deeply contested, as it is one in which private sector interests have been particularly vocal and where the voices of vulnerable people and communities are crying to be heard.

But the biodiversity convention, like all the resolutions, declarations and treaties, can and must be shaped to serve the needs and values of their people, not the interests of transnational corporations and rapacious asset managers. We badly need governments to come together with civil society to save this gift of a planet. The time for leadership has never been greater.

AFTERWORD

This was a difficult book to write, as I am sure it was to read. The last book I wrote, *Still Hopeful: Lessons from a Lifetime of Activism*, was published in 2022. So much has happened since then, so much of it bad, that many people ask me if I still have hope.

My answer is yes. I believe with all my heart that hope is a moral imperative. To give up on hope is to give up on the young, and that would be profoundly wrong. Others have gone through terrible times before us but have survived and built solid lives founded on solid values. In understanding the role of hope, I borrowed a term coined by American spiritual advisor Roshi Joan Halifax, who counsels men on death row.

Halifax uses the term "wise hope," and I have interpreted it this way: Wise hope requires that you look at the situation squarely and not sugar-coat it. The more you can learn, the better, no matter how painful. In spite of that painful knowledge, you take action. You put your hand out to touch the universe in a way that you can affect it, however small that may feel to you. But — and this is important — you must have faith that many others around the world are doing the same, even though you cannot know who they all are or what they are contributing. Hope requires action; action requires hope. Both require faith in a larger good.

My husband, Andrew, and I have four young adult grandchildren, and their very presence in the world is cause for hope. They and their peers are leading the way and give me grace and solace. I hope with all my being that this book — a cry from the heart — causes many to see the folly of trusting the care of the Earth to the very corporations and faceless hedge funds that have caused such destruction. In a world where two-thirds of the climate damage is created by just 10% of the population, and where a new cohort of trillionaires is about to walk among us, something is profoundly wrong. We need systemic analysis and systemic change to take care of our natural heritage and one another.

Meanwhile, every time you do something positive — plant a tree, stand up to an aggressor, fight for better law, help someone more vulnerable, speak up against racism, protect a lake, help the young — you are the antidote to the greed, violence and injustice in our world. We can walk this path together.

> *Nature has a lot to say, and it has long been time for us, her children, to stop playing deaf. Maybe even God will hear the cry rising from this Andean country and add an eleventh amendment, which he left out when he handed down instructions from Mount Sinai: "Love Nature, which you are a part of."*
>
> — EDUARDO GALEANO, LATE URUGUAYAN WRITER AND HUMAN RIGHTS ACTIVIST

ACKNOWLEDGEMENTS

This book was inspired by an invitation to write a discussion paper for the Global Alliance for the Rights of Nature on how the financialization of nature is impacting this movement. I am deeply grateful to Natalia Greene, Cormac Cullinan and Shannon Biggs for their faith in me.

Frederic Hache of Green Finance Observatory has been a fountain of information and advice, and I am very grateful to him for his life's work. David Boyd, Canadian legal expert on human rights and the environment, has been a true guide in understanding rights-based law.

I thank Ellen Dorsey, Owen Mutiganda and the team at Wallace Global Fund for financial support in the research and writing of this book.

Once again I am so grateful to David Caron and ECW for their unwavering support in undertaking yet another book from me! As always, I cannot thank editor Susan Renouf enough for her expertise and for making this book, as she has done with all the others, better. Jen Knoch, as well, did a fabulous job of copy editing (and a bit more). Thank you all!

My husband, Andrew, and my family have been fully supportive — if a little surprised — that my "retirement" includes writing such challenging books. I am deeply grateful to them for their love.

NOTES

1. Dieter Helm, *The State of Natural Capital: Restoring Our Natural Assets*, Natural Capital Committee (UK), March 2014, online.
2. Kelli Meskin Soileau, "Whales Are Now Legal Persons," *Ka Wai Ola*, June 1, 2024, online.
3. Fabio Berzaghi et al., "Value Wild Animals' Carbon Services to Fill the Biodiversity Financing Gap," *Nature Climate Change* 12 (2022): 598–601, online.
4. Adrienne Buller, *The Value of a Whale: On the Illusion of Green Capitalism* (Manchester University Press, 2022).
5. George Monbiot, "The UK Government Wants to Put a Price on Nature — but That Will Destroy It," *The Guardian*, May 15, 2018, online.
6. Antonio Tricarico and Caterina Amicucci, *Financialization of Water* (Alternative Water Forum, December 16, 2012).
7. Seth Borenstein, "The World's Biggest Companies Have Caused $28 Trillion in Climate Damage, a New Study Estimates," *The Globe and Mail*, April 23, 2025, online.
8. William E. Rees, "The Human Ecology of Overshoot,"*MDPI*, August 11, 2023, online.
9. *State of Finance for Nature 2023* (UN Environment Programme and partners, December 9, 2023), online.
10. Richard Bocking, "Reclaiming the Commons," Canadian Unitarian Council newsletter, 2023.

11. Jonathan Rowe, "Fanfare for the Commons: The Trees, the Air, Open Spaces, the Sky," *Utne Reader* 109 (2002).
12. Alexander Koch et al., "Earth System Impacts of the European Arrival and Great Dying in the Americas After 1492," *Quaternary Science Reviews*, March 1, 2019, online.
13. National Consumers League, "Legal Victory for Consumers: NCL Lawsuit Against Starbucks Moves Forward," press release, August 11, 2025, online.
14. Nicolas McMillen, *Gold and Shadows: The Illicit Plunder of Latin America's Resources* (American Bar Association, August 2, 2024), online.
15. Joel Wendland-Liu, "New Study Investigates Plunder of Native American Land," *People's World*, June 13, 2022, online.
16. Palko Karasz, "Half of England Is Owned by Less Than 1% of Its Population, Researcher Says," *New York Times*, April 19, 2019, online.
17. Jason Hickel, "How Britain Stole $45 Trillion from India," *Al Jazeera*, December 19, 2018, online.
18. Vandana Shiva et al., *The Enclosure and Recovery of the Commons* (Research Foundation for Science, Technology, and Ecology, 1997).
19. Taiwo Adebayo, "Activists Urge Nigeria to Delay Shell's $2.4 Billion Sale of Assets in Deeply Polluted Niger Delta," Associated Press, February 28, 2024, online.
20. Amnesty International, "Nigeria: Shell Must Be Held Fully Accountable for Human Rights Harms Before Being Allowed to Sell Its Niger Delta Business," press release, May 14, 2024, online.
21. Jann Lay et al., *Taking Stock of the Global Land Rush* (Centre for Development and Environment, University of Bern, 2021), online.
22. Jason Hickel et al., "Rich Countries Drained $152 Tn from the Global South Since 1960," *Al Jazeera*, May 5, 2021, online.

23. Steven Bernstein, *The Compromise of Liberal Environmentalism* (Columbia University Press, 2001).
24. The Bretton Woods Project, *What Are the Main Criticisms of the World Bank and the IMF?*, June 4, 2019, online.
25. John Cavanagh and Jerry Mander, "World Bank, IMF Turned Poor Third World Nations into Loan Addicts," *The CCPA Monitor*, July 2003, online.
26. Brian Roach, *Corporate Power in a Global Economy* (Global Development Policy Center, Boston University, 2023), online.
27. Global Justice Now, "69 of the Richest 100 Entities on the Planet Are Corporations, Not Governments, Figures Show," October 17, 2018, online.
28. Research FDI, "Everything You Need to Know About Trade Agreements," Montreal, December 9, 2022, online.
29. Vandana Shiva et al., *The Enclosure and Recovery of the Commons*.
30. Fiona Harvey, "Fossil Fuel Firms Forcing Countries to Compensate Them, Mary Robinson Says," *The Guardian*, May 14, 2024, online.
31. Phoebe Weston and Patrick Greenfield, "Why Fear of Billion-Dollar Lawsuits Stops Countries Phasing Out Fossil Fuels," *The Guardian*, March 6, 2025, online.
32. Canadian Centre for Policy Alternatives, "Canadian Investor-State Lawsuits from Mining Companies Threaten Environmental Policy and Sustainable Development: Study," press release, May 19, 2022, online.
33. Transnational Institute, *The Financialisation of International Investment Law*, March 26, 2025, online.
34. Patrick Greenfield and Phoebe Weston, "Revealed: How Wall Street Is Making Millions Betting Against Green Laws," *The Guardian*, March 5, 2025, online.
35. Katie Surma and Nicholas Kusnetz, "How to Buy a Piece of a Lawsuit and Impoverish a Country," *Inside Climate News*, December 8, 2024, online.

36. M. Ayhan Kose and Alen Mulabdic, "Global Trade Has Nearly Flatlined. Populism Is Taking a Toll on Growth," *World Bank Blog*, February 22, 2024, online.
37. Farah Stockman, "The WTO Is Having a Midlife Crisis," *New York Times*, December 17, 2020, online.
38. Paul Krugman, "What Economists (Including Me) Got Wrong About Globalization," *Bloomberg*, October 10, 2019, online.
39. Michael Smolyansky, *End of an Era: The Coming Long-Run Slowdown in Corporate Profit Growth and Stock Returns* (Board of Governors of the Federal Reserve System, 2023), online.
40. Patrick Greenfield, "Investing 0.1% of Global GDP Could Avoid Breakdown of Ecosystems, Says UN Report," *The Guardian*, May 27, 2022, online.
41. Patrick Greenfield and Phoebe Weston, "'Two Sides of the Same Coin': Governments Stress Links Between Climate and Nature Collapse," *The Guardian*, November 4, 2024, online.
42. Phoebe Weston, "Land Could Be Worth More Left to Nature Than When Farmed, Study Finds," *The Guardian*, March 8, 2021, online.
43. Rochelle Baker, "Old-Growth in Contentious Fairy Creek Region Could Be Worth More Standing Than Logged," *Canada's National Observer*, June 30, 2021, online.
44. Sonal Gupta, "BC's Interior Old-Growth Forests Hiding Billions in Economic Benefits, Report Says," *Canada's National Observer*, April 14, 2025, online.
45. "Nature-Based Solutions," David Suzuki Foundation, accessed September 2, 2025, online.
46. "Nature-Based Solutions," Directorate-General for Research and Innovation, European Commission, accessed September 2, 2025, online.
47. Brenden Jongman and Sally Judson, "Nature-Based Solutions for Climate Resilience Are Catching on in World Bank Projects," *World Bank blog*, March 29, 2023, online.

48. United Nations Environment Programme, *Protected Planet Report 2024*, October 2024, online.
49. *Don't Bet on Wall Street: The Financialization of Nature and the Risk to our Common Resources* (Food & Water Watch, June 2012).
50. Robert Costanza et al., "The Value of the World's Ecosystem Services and Natural Capital," *Nature* 387 (May 15, 1997), online.
51. William K. Stevens, "How Much Is Nature Worth? For You, $33 Trillion," *New York Times*, May 20, 1997, online.
52. Walter V. Reid et al., *Millennium Ecosystem Assessment: Ecosystems and Human Well-Being* (World Resources Institute, March 1, 2005), online.
53. *The New Business Imperative: Valuing Natural Capital* (Corporate EcoForum and the Nature Conservancy, 2012), online.
54. *Living Planet Report, 2018: Aiming Higher* (WWF, 2018), online.
55. *The High Cost of Cheap Water: The True Value of Water and Freshwater Ecosystems to People and Planet* (World Wildlife Fund, 2023), online.
56. *The New Business Imperative.*
57. "United Nations Carbon Offset Platform," United Nations, accessed September 2, 2025, online.
58. *World Bank Group and the Global Environment Facility* (World Bank, August 11, 2023), online.
59. "Investing in Nature Unlocks Development Benefits," World Bank Group, August 5, 2021, online.
60. Oxfam International, "Up to $41 Billion in World Bank Climate Finance Unaccounted for, Oxfam Finds," press release, October 17, 2024, online.
61. Aubrey R. Paris, "Science Speaks: Green Economy," US Department of State, September 30, 2021, online.
62. Rob Soutar, "What Trump's Second Term Means for Sustainable Finance," *The Bureau of Investigative Journalism*, January 16, 2025, online.

63. Max Fawcett, "Donald Trump Can't Slow Down the Energy Transition," *Canada's National Observer*, February 14, 2025, online.
64. World Bank, "State of California Invests in World Bank 5-Year Green Bond," press release, April 9, 2024, online.
65. *Australian Green Economy Prospectus: A Global Transformation Towards a Green Economy* (Australian Government, 2023), online.
66. UN Environment Programme, "Private Finance for Nature Surges to $102 Billion," June 10, 2024, online.
67. Adrien Sénécat, "The Great Green Investment Investigation," *Le Monde*, November 29, 2022, online.
68. Josh Gabbatiss, "Analysis: How Some of the World's Largest Companies Rely on Carbon Offsets to 'Reach Net-Zero,'" *Carbon Brief*, September 27, 2023, online.
69. "Global Carbon Offset Market Set to Soar to US$1.6 Trillion By 2028," DGB Group, October 11, 2023, online.
70. "Carbon Offset Market Growth, Trends and Forecasts to 2032," *Yahoo Finance*, January 28, 2025, online.
71. "320 Companies and Financial Institutions to Start TNFD Nature-Related Corporate Reporting," Taskforce on Nature-Related Financial Disclosers, January 16, 2024, online.
72. Timothy Workman, "Is TFND an Answer to the Biodiversity Crisis?" *Rainforest Action Network Blog*, September 23, 2025, online.
73. "Business for Nature, More Than 130 Companies and Financial Institutions Call for Renewed Policy Ambition to Implement the Biodiversity Plan," press release, July 15, 2024, online.
74. Lydia Larsen, "Just Five Companies Produce Nearly 25 Percent of All Plastic Waste Worldwide," *Sierra Magazine*, May 20, 2024, online.
75. John Bellamy Foster, "The Defense of Nature: Resisting the Financialization of the Earth," *Monthly Review*, April 1, 2022, online.

76. Kate Mackenzie, "Uneven Channels, Climate Diplomacy and the Global Financial Architecture," *Phenomenal World*, October 30, 2021, online.
77. Thelma Arko, "Carbon Markets and the New Scramble for African Land," *Progressive International*, January 2, 2025, online.
78. Jutta Kill, *Financialisation of Nature: Creating a New Definition of Nature* (Friends of the Environment, 2015), online.
79. Ed Davey, "In Peru, Kichwa Tribe Wants Compensation for Carbon Credits," Associated Press, December 22, 2022, online.
80. Buller, *The Value of a Whale*.
81. George Monbiot, "Put a Price on Nature? We Must Stop This Neoliberal Road to Ruin," *The Guardian*, July 24, 2014, online.
82. Nature Canada, "Poll: Canadians Want Parties to Prioritize Nature in Federal Election," press release, November 27, 2024, online.
83. Alec Tyson and Brian Kennedy, "Two-Thirds of Americans Think Government Should Do More on Climate," *Pew Research Center*, June 23, 2020, online.
84. Dylan Difford, "How Do Britons Feel About Expanding Renewable Energy?" *YouGov*, August 12, 2024, online.
85. YouGov, "YouGov Survey: Seriousness of National Problems," August 4–7, 2025, online.
86. *Europeans' Attitudes Towards EU Energy Policy* (European Commission, 2024), online.
87. European Investment Bank, "Nearly 9 Chileans in 10 Demand Stricter Climate Policies, EIB Survey Reveals," press release, September 4, 2023, online; European Investment Bank, "8 Brazilians in 10 Demand Stricter Climate Policies, EIB Survey Reveals," press release, September 4, 2023, online.
88. "Most Indians Want Stronger Climate Action, Survey Says," *The Hindu*, June 24, 2024, online.

89. "80 Percent of People Globally Want Stronger Climate Action by Governments According to New Survey," University of Oxford, June 20, 2024, online.
90. Damian Carrington, "Activate Climate's 'Silent Majority' to Supercharge Action, Experts Say," *Canada's National Observer*, April 22, 2025, online.
91. Buller, *The Value of a Whale*.
92. Bloomberg Intelligence, "ESG AUM Set to Top $40 Million by 2030, Anchor Capital Markets," *Bloomberg Professional Services*, February 22, 2024, online.
93. Derek Seidman, "Can We Overcome the Asset Manager Economy?" *Nonprofit Quarterly*, March 6, 2024, online.
94. Buller, *The Value of a Whale*.
95. *Executive Perspectives on Top Risks for 2024–2034* (Proviti, 2024), online.
96. Buller, *The Value of a Whale*.
97. Andre Standing, "The Financialization of Conservation," Transnational Institute, November 21, 2022, online.
98. Hugh Possingham, "The New Mayor Players in Conservation? NGOs Thrive While National Parks Struggle," *The Conversation*, February 26, 2023, online.
99. Zack Colman, "'The System Was Broken': How the Nature Conservancy Prospered but Ran Aground," *Politico*, July 7, 2019, online.
100. "NatureVest," The Nature Conservancy, accessed September 2, 2025, online.
101. Kenza Bryan, "'Sustainable' Debt Pioneer Ditches Controversial 'Blue Bond' Label," *Financial Times*, September 22, 2023, online.
102. John Vidal, "WWF International Accused of 'Selling Its Soul' to Corporations," *The Guardian*, October 4, 2014, online.
103. "Nature-Based Solutions Origination Platform," World Wildlife Fund, accessed September 2, 2025, online.
104. "Natural Capital Accounting," European Commission, accessed September 2, 2025, online.

105. "Natural Capital Accounting Community of Practice Africa," accessed September 2, 2025, online.
106. US White House, "A New National Strategy to Reflect Natural Assets on America's Balance Sheet," press release, August 18, 2022, online.
107. "About the Census of Environment," Statistics Canada, accessed September 2, 2025, online.
108. Accounting for Nature, accessed September 2, 2025, online.
109. "Partnership on Cooperation on Nature Capital Accounting, Environmental-Economic Accounting, and Related Statistics," Government of Canada, accessed September 2, 2025, online.
110. Stanford Natural Capital Project, "Gross Ecosystem Product (GEP) (China)," accessed September 2, 2025, online.
111. "The Green Bond Market, Currently at $4 Trillion, Needs to Expand Tenfold to Meet the Urgent Demands of Climate Action," Climate Bonds Initiative, March 16, 2024.
112. "Are Green Bonds as Good as They Sound?" *Berkeley Economic Review*, November 26, 2019, online.
113. "New Investigation Alleges Deforestation and Greenwashing Linked to Michelin," *Mighty Earth News*, November 15, 2022, online.
114. Joel Makower, "Why ESG Will Still Be the Wild West In 2022," *Trellis*, November 29, 2021, updated July 24, 2024, online.
115. Stefano Valentino and Giorgio Michalopoulos, "How Big Finance Greenwashes Climate Crisis Culprits," *Voxeurop*, June 25, 2024, online.
116. Matt Macfarland et al., "Green Funds Have a Paris Alignment Problem," S&P Global, June 7, 2022, online.
117. Ishika Mookerjee, "Green Bonds Aren't Driving New Climate Action in US, Study Says," *Bloomberg*, September 19, 2024, online.
118. Jutta Kill, "REDD: The Pitfalls of Market-Compliant Conservation," Heinrich Böll-Stiftung, August 28, 2020, online.

119. "Mozambique's Pulp Fiction," *Water Journalists Africa*, April 24, 2025, online.
120. "What the Company Leaves for the Population is 0%," World Rainforest Movement, October 24, 2024, online.
121. "Reforestation," Green Earth Africa, accessed September 2, 2025, online.
122. Fiona Harvey, "Global Carbon Trading System Has 'Essentially Collapsed,'" *The Guardian*, September 10, 2012, online.
123. Chloé Farand, "'Hot Air' Carbon Offset Scheme Undermines Colombia's Climate Goal, Experts Warn," *Climate Home News*, June 30, 2021, online.
124. John Cannon, "UN Probes Controversial Forest Carbon Agreement in Malaysian Borneo," *Mongabay*, November 9, 2021, online.
125. Sam Quashie-Idun and Emma Howard, "'How Are We Going to Live?' Families Dispossessed of Their Land to Make Way for Total's Congo Offsetting Project," *Unearthed, Greenpeace*, December 12, 2022, online.
126. *Selling Hot Air: Lessons From How Shell's Flagship Carbon Capture Project Sold $200M of Credits for Reductions That Never Happened* (Greenpeace Canada, February 2024), online.
127. *Up in Smoke: Wildfires and the Myth of Carbon Offsets* (Food & Water Watch, May 2024), online.
128. Patrick Greenfield, "The New 'Scramble For Africa': How a UAE Sheikh Quietly Made Carbon Deals for Forests Bigger Than UK," *The Guardian*, November 30, 2023, online.
129. Nina Lakhani, "Revealed: Top Carbon Offset Projects May Not Cut Planet-Heating Emissions," *The Guardian*, September 19, 2023, online.
130. Adam Morton, "Australia's Carbon Credit System a Failure on Global Scale, Study Says," *The Guardian*, March 26, 2024, online.

131. Climate Analytics, "Why Offsets Don't Work: New Analysis," press release, February 10, 2023, online.
132. Elina Kajosaari, "Community Conflict and Vague Predictions: The Five Biggest Reasons Carbon Offsetting Schemes Fail," *Euronews*, June 1, 2021, online.
133. "Elon Musk Earns Record Amount from Carbon Trading," *Follow the Money*, January 30, 2025, online.
134. Graham Readfearn, "Corporations Using 'Ineffectual' Carbon Offsets Are Slowing Path to 'Real Zero,' More Than 60 Climate Scientists Say," *The Guardian*, October 27, 2024, online.
135. Shaina Sadai, "Fossil Fuel Companies Make Billions in Profit as We Suffer Billions in Losses: 2024 Edition," Union of Concerned Scientists, April 17, 2024, online.
136. Ben McKay, "Climate Advocates Label Australia a 'Petro-State,'" *The Canberra Times*, March 14, 2025, online.
137. Zanagee Artis, "Big Oil Made Billions Amid the Hottest Year on Record," NRDC, February 29, 2024, online.
138. Gujari Singh, "Big Oil Spends Millions to Block Climate Action in California — Waging War on Californians," Greenpeace, February 4, 2025, online.
139. Simon Black et al., *IMF Fossil Fuel Subsidies Data: 2023 Update* (IMF, August 2023), online.
140. Julia Levin, *Canada's Fossil Fuel Funding in 2023* (Environmental Defence, March 2024), online.
141. Jeff Lagerquist, "Canada's Aid to the Oil & Gas Sector Topped $29.6B in 2024: Environmental Defence," *Yahoo Finance*, April 3, 2025, online.
142. John Woodside, "Oil and Gas Companies Plan to Spend Billions to Boost Future Fossil Fuel Supply," *Canada's National Observer*, March 11, 2025, online.
143. Megan He et al., "Total Organic Carbon Measurements Reveal Major Gaps in Petrochemical Emissions Reporting," *Science* 383, no. 6681 (January 25, 2024), online.

144. Chris Lang, "'We Must Say NO! to This Green Economy Regime.' Indigenous Environmental Network Calls for a Moratorium on all False Solutions to the Climate Crisis," *REDD-Monitor*, May 17, 2025, online.
145. Anusha Krishnan, "What Is Biodiversity Finance?" *Mongabay*, September 13, 2022, online.
146. "A Way Forward for High Integrity Biodiversity Credits," International Advisory Panel on Biodiversity Credits, October 28, 2024, online.
147. Aruna Chandrasekhar, "In-Depth Q&A: What Are 'Biodiversity Offsets'?" *Carbon Brief*, September 26, 2023, online.
148. "New IPBES Reports Reveal Huge Economic and Business Opportunities for Accelerating Biodiversity Action," European Union, December 28, 2024, online.
149. NatureFinance, accessed September 21, 2025, online.
150. NatureFinance, accessed September 2, 2025, online.
151. "On Behalf of HSBC Global Asset Management CEO Nicolas Moreau Pledges for Biodiversity," Finance for Biodiversity, September 25, 2020, YouTube, online.
152. "Goldman Sachs AM Launches First Biodiversity Bond Fund, Targeting Up to $500 Million," ESG News, February 28, 2025, online.
153. High Ambition Coalition, accessed September 2, 2025, online.
154. UN Environment Programme, "World Must Act Faster to Protect 30% of the Planet by 2030," press release, October 28, 2024, online.
155. Josh Davis, "Can Privatization Save Parks?" *Mongabay*, February 26, 2016, online.
156. Robert Williams, "Kicking Native People off Their Land Is a Horrible Way to Save the Planet," *New York Times*, February 20, 2024, online.
157. "Biodiversity's Greatest Protectors Need Protection," *Scientific American*, October 1, 2021, online.

158. Oakland Institute, "From Abuse to Power: Ending Fortress Conservation in the Democratic Republic of Congo," press release, August 27, 2024, online.
159. "Blood Carbon," Survival International, January 26, 2025, online.
160. Fiona Stewart, Samantha Power and Shaun Mann, "Harnessing the Power of Capital Markets to Conserve and Restore Global Biodiversity Through 'Natural Asset Companies,'" *World Bank Blog*, October 12, 2021, online.
161. Declan Harty, "NYSE's New Investment Vehicle — 'Natural Asset Companies' — Will Tap into ESG Fever," *Fortune*, September 14, 2021, online.
162. Center for American Progress, "Congress' Tax Bill Is Selling Out America's Public Lands and Waters," press release, May 6, 2025, online.
163. Wild Capital, accessed September 2, 2025, online.
164. William Booth, "In Scotland 'Green Lairds' Are Buying Vast Estates in Scotland for Carbon Credits," *The Washington Post*, February 15, 2025, online.
165. Lisa Cox, "'Utterly Damning' Review Finds Offsets Scheme Fails to Protect NSW Environment," *The Guardian*, August 30, 2022, online.
166. Penny van Oosterzee and Jayden Engert, "As Australia Privatises Nature Repair, the Cheapest Approach Won't Save Our Threatened Species," *The Conversation*, November 17, 2024, online.
167. Greenpeace, "Greenpeace Canada Comments — Offsetting Policy for Biodiversity," February 17, 2023, online.
168. Sarah Sobanski, "Investors Keen to Grow Nature Conservation Investment in Canada," *Carbon Pulse*, February 25, 2025, online.
169. Vandana Shiva, "Biodiversity Is Life, Not an Asset Class," Navdayana International, accessed September 24, 2025, online.
170. Steve Stawarz, "Biodiversity Credits and Conservation," The Conservation Foundation, February 12, 2024, online.

171. "We Are Not Here to Renegotiate the Convention, Says IPC," transcript of presentation published by La Via Campesina, December 21, 2022.
172. Frederic Hache, "Can Finance Save Polar Bears?" *The Mint Magazine*, March 26, 2024, online.
173. "Don't Be Fooled! Biodiversity Markets Are False Solutions," civil society statement on biodiversity offsets and credits, Biomarketwatch.info, September 2024, online.
174. United Nations Environment Programme, "Plastic Pollution," accessed September 25, 2025, online.
175. Andrew Blok, "Record Levels of Harmful Particles Found in Great Lakes Fish," *Environmental Health News*, July 29, 2021, online.
176. Marcus Eriksen et al., "A Growing Plastic Smog, Now Estimated to Be Over 170 Trillion Plastic Particles Afloat in the World's Oceans — Urgent Solutions Required," *PLOS One* (March 8, 2023), online.
177. Fermin Koop, "Amazon's Plastic Packaging Waste Could Encircle the Globe 500 Times," *ZME Science*, December 16, 2020, online.
178. Stefanie Valentic, "Plastics Manufacturing and Greenhouse Gas Emissions: Are Plastics the New Coal?" Beyond Plastics, October 26, 2021, online.
179. Sandee LaMotte, "Toxic Plastic Chemicals Number in the Thousands, Most Are Unregulated, Report Finds," *CNN*, March 14, 2024, online.
180. Verner Viisainen and Ayesha Tandon, "Five Charts: Why a UN Plastics Treaty Matters for Climate Change," *Carbon Brief*, November 25, 2024, online.
181. Rachel Ramirez, "The Plastic Water Bottle Industry Is Booming. Here's Why That's a Huge Problem," *CNN*, March 16, 2023, online.
182. Issam Ahmed, "Over a Billion Pounds of Coke Plastic Waste Enter Waterways: Study," *Phys.org*, March 26, 2025, online.

183. Peter Gleick, "Bottled Water and Energy: A Pacific Institute Fact Sheet," Pacific Institute, 2007, online.
184. Carlos Mureithi and Jennifer Mcdermott, "What on Earth Are Plastic Credits and Will They Really Reduce Plastics Pollution?" *Canada's National Observer*, November 17, 2023, online.
185. *Plastic Credits at a Glance* (World Bank, November 2024), online.
186. Kate Aronoff, "The Plastic Industry's Latest Delay Tactic: 'Plastic Offsets,'" *The New Republic*, May 3, 2024, online.
187. Rémy Bourdillon, "A Canadian Company Says It's Fighting Pollution in the Philippines. Is it Cashing in Instead?" *The Walrus*, April 8, 2025, online.
188. Emma Howard, "Bali Plastic Recycling Scheme Swamped with Garbage," *The Guardian*, November 20, 2024, online.
189. Thalia Bofiliou et al., *Alliance to End Plastic Waste: Barely Credible* (Planet Tracker, August 2022), online.
190. Sandra Laville, "Five Firms in Plastic Pollution Alliance 'Made 1,000 Times More Plastic Than They Cleaned Up,'" *The Guardian*, November 20, 2024, online.
191. IUCN, "The Coca-Cola Foundation donates $10 Million to Help Reduce Plastic Packaging Waste in 7 African Islands Through IslandPlas Project," press release, June 12, 2024, online.
192. Center for International Environmental Law, "Fossil Fuel Lobbyists Outnumber National Delegations, Scientists, and Indigenous Peoples at Plastics Treaty Negotiations," press release, April 25, 2024, online.
193. Damian Carrington, "'Total Infiltration': How Plastics Industry Swamped Vital Global Treaty Talks," *The Guardian*, July 23, 2025, online.
194. Frederic Hache, *Offsets On-Chained* (Green Finance Observatory, 2022), online.
195. Marco Schletz et al., "Blockchain and Regenerative Finance: Charting a Path Toward Regeneration," *Frontiers* 6 (2023), online.

196. Hache, *Offsets On-Chained*.

197. "Blockchain and Smart Contracts: Capital's Latest Attempt to Seize Life on Earth," World Rainforest Movement, January 13, 2020, online.

198. *Digital Economy Report: Shaping an Environmentally Sustainable and Inclusive Digital Future* (United Nations, 2024), online.

199. Hache, *Offsets On-Chained*.

200. Tom Gerken, "Argentina President Faces Impeachment Calls Over Crypto Crash," *BBC*, February 17, 2025, online.

201. Daniel de Visé, "Republicans Loved Crypto Before Trump Jumped on the Bandwagon. Here's Why," *USA Today*, January 7, 2025, online.

202. Vicky Ge Huang, "The Trump Family Advances Its All-Out Crypto Blitz, This Time with Bitcoin Mining," *Wall Street Journal*, March 31, 2025, online.

203. Angus Berwick, "Trump Family Amasses $5 Billion Fortune After Crypto Launch," *Wall Street Journal*, September 1, 2025, online.

204. Chris Gay, "Trump's Weird Crypto Obsession Could Spark the Next Financial Crisis," *Globe and Mail*, March 19, 2025, online.

205. Peter Waldman et al., "The Water Trade Is Booming — and Sucking Australia Dry," *Bloomberg*, December 27, 2023, online.

206. Quentin Grafton et al., "Declining Streamflow in the Northern Murray-Darling Basin: It's Not Just Climate Change," Global Water Forum, March 27, 2025, online.

207. Waldman et al., "The Water Trade Is Booming."

208. Jeremy Rifkin, *Planet Aqua: Rethinking Our Home in the Universe* (Polity Press, 2024).

209. Kyle Frankel Davis et al., "Water Limits to Closing Yield Gaps," *Advances in Water Resources* 99 (January 2017), online.

210. Dongmei Feng and Colin J. Gleason, "More Flow Upstream and Less Flow Downstream: The Changing Form and

Function of Global Rivers," *Science* 386, no. 6727 (December 12, 2024), online.

211. Jet Propulsion Laboratory, "NASA Satellites Reveal Abrupt Drop in Global Freshwater Levels," press release, November 19, 2024, online.
212. "Over Half of World's Largest Lakes Dwindle, Threatening Water Security," *Thaiger*, May 19, 2023, online.
213. Lisa Gross, "Groundwater Levels Around the World Are Dropping Quickly, Often at Accelerating Rates," *Inside Climate News*, January 24, 2024, online.
214. Fiona Harvey, "Global Fresh Water Demand Will Outstrip Supply by 40% by 2030, Say Experts," *The Guardian*, March 17, 2023, online.
215. World Health Organization, "1 in 3 People Globally Do Not Have Access to Safe Drinking Water — UNICEF, WHO," press release, June 18, 2019, online.
216. Crawford Kilian, "When Cities Hit 'Day Zero' with No Water," *The Tyee*, May 23, 2024, online.
217. IUCN, "One Quarter of Freshwater Animals at Risk of Extinction — IUCN Red List," press release, January 8, 2025, online.
218. "Evolution des modes de gestion de l'eau par département (régies publiques, délégations)," Akwari Coop, accessed September 25, 2025, online.
219. Jessica Shankleman, "Why Customers Will Pay the Bill for UK's Water Crisis," *BNN Bloomberg*, December 20, 2024, online.
220. Yasmin Rufo, "Thames Water: 72 Billion Litres of Sewage Pumped into River in Two Years," *BBC*, November 10, 2023, online.
221. Camilla Turner, "Every River in the Country Is Polluted, Environment Secretary Warns," *The Telegraph*, July 20, 2024, online.
222. David Parsley, "More Than Half of Water Bills in England

Help Fund Chinese Investment Firms," *The i Paper*, March 13, 2023, online.

223. Philip Alston, *Extreme Poverty and Human Rights* (United Nations, 2018), online.
224. "Veolia and Suez: the New Face of the Private Water Giants," Public Services International, March 22, 2022, online.
225. Neil Gupta, "Veolia and Suez Merger Directly Threatens Human Right to Water," *Corporate Accountability*, June 30, 2021, online.
226. "About GCF," Green Climate Fund, accessed September 2, 2025, online.
227. Gupta, "Veolia and Suez Merger."
228. Sara Larraín and Colombina Schaeffer, *Conflicts over Water in Chile: Between Human Rights and Market Rules* (Chile Sustenable, 2010), online.
229. Center for Climate and Resilience Research, "Water Crisis in Chile: Are We Close to Day Zero?" April 30, 2024, online.
230. Kara Anderson, "The Harmful Effects of Our Lithium Batteries," *Greenly*, July 23, 2024, online.
231. John Bartlett, "How One Community in Chile Is Blessed and Cursed with Lithium," *NPR*, February 23, 2025, online.
232. Nidhi Sharma, "Water Shortage Threatens the World's Most Abundant Lithium Reserves," *NBC News*, March 31, 2025, online.
233. Leslie Kaufman et al., "Arizona Is Running Out of Cheap Water. Investors Saw It Coming," *Bloomberg*, June 21, 2023, online.
234. Katie Rose Quandt, "Corporations Plunder US West's Water Amid Worst Drought in 1,200 Years," *Truthout*, September 22, 2024, online.
235. Peter Waldman, Momar Niang and Katarina Hoije, "A Global Hunt for Water Profit Risks Draining Cities Dry," *Bloomberg*, November 14, 2023, online.

236. Peter Waldman et al., "Groundwater Gold Rush," *Bloomberg*, April 11, 2023, online.
237. Dina Arévalo, "SpaceX Consumes Millions of Gallons of Brownsville Water Amid Scarcity," *My RGV*, November 22, 2024, online.
238. Andrew Collier, "Artificial Intelligence Is Using a Ton of Water. Here's How to Be More Resourceful," Veolia, August 22, 2024, online.
239. Luke Barratt, "Revealed: Big Tech's New Datacentres Will Take Water from the World's Driest Areas," *The Guardian*, April 9, 2025, online.
240. Leonardo Nicoletti, Michelle Ma and Dina Bass, "AI Is Draining Water from Areas That Need It Most," *Bloomberg*, May 8, 2025, online.
241. Omar Ocampo and A.J. Schumann, "AI's Energy Demands Are Fueling the Climate Crisis," *Inequality.org*, December 16, 2024, online.
242. Amanda Woodard, "Water Security: Accounting for Water," *INTHEBLACK*, November 1, 2023, online.
243. Brian Richter, "Water Markets Can Support an Improved Water Future," CEO Water Mandate, August 2016, online.
244. Michael Muir, "Scientists Sound Alarm on Hidden Threat Lurking Within Tap Water System: 'This Is a Call to Action.'" *The Cool Down*, March 11, 2025, online.
245. Nestlé, "Nestlé Group Reports Solid 2024 Performance; Executing on Plan to Accelerate Growth," press release, February 13, 2025, online.
246. Isaac Lane, "Nestlé's $12 Billion Water Division Sale: A Strategic Shift with High-Reward Risks," *AInvest*, May 10, 2025, online.
247. Amos Abba, "How Nestle Nigeria Contaminates Water Supply of Its Host Community in Abuja," *International Center for Investigative Reporting*, April 28, 2019, online.

248. "Cancel & Prevent Transfer of Nestlé's Water-Taking Permits," Wellington Water Watchers, accessed September 2, 2025, online.
249. Wellington Water Watchers, "Water Watchers Oppose Move to Make Water Taking Permits Transferable," press release, July 10, 2025, online.
250. "China's Richest Person Made Billions Bottling Pristine Water," *Bloomberg*, October 31, 2023, online.
251. Grace Livingstone, "'It's Not Drought — It's Looting': The Spanish Villages Where People Are Forced to Buy Back Their Own Drinking Water," *The Guardian*, November 23, 2024, online.
252. Peter Yeung, "'Forgotten': How One Mexican City Struggles Against Big Industry for Water," *Al Jazeera*, January 8, 2025, online.
253. "About Us," MarillionEx, accessed September 2, 2025, online.
254. "Water Offsets & Credits," Green Earth Africa, accessed September 2, 2025, online.
255. *BioCarbon Water Standard: Sustainable Management and Conservation of Water Resources* (BioCarbon, 2025), online.
256. Christine MacDonald, "How Coke Spun the Public on Its Water Use," *Type Investigations*, May 31, 2018, online.
257. Break Free from Plastics, "Coca-Cola Erases Reusable Targets," press release, December 4, 2024, online.
258. "Trading and Offsets in the Chesapeake Bay Watershed," United States Environmental Protection Agency, March 15, 2024, online.
259. "Water Quality Trading," World Resources Institute, accessed September 2, 2025, online.
260. *Water Quality Trading: Polluting Public Waterways for Private Gain* (Food & Water Watch, November 2015), online.
261. Alexander M. Ervin et al., eds., *Corporate Hog Barns and the Threat to Public Health, the Environment and Rural Communities* (Canadian Centre for Policy Alternatives, 2003), online.

262. *Water Security and Climate Change: Insights from Country Climate and Development Reports* (World Bank Group, March 4, 2025), online.
263. Evelyn Balassiano and John Guzman, "Under Pressure: Latin American Water Sector Turns to Private Investment," White & Case, November 18, 2024, online.
264. Hervé Thiard, "Bourse: 'L'eau est un bon placement, qui pourrait rapporter 8% par an,'" *Capital*, February 2, 2024, online.
265. Susannah Savage, "Water Demand Will Soar Because of Wars and Tech, Says Utility Boss," *Financial Times*, October 7, 2024, online.
266. "Marine Water Desalination Industry Research 2024: Market to Almost Double by 2028 with Significant Growth Projected, Reaching $36.67 Billion by 2033," GlobeNewswire, January 3, 2025, online.
267. "We Cannot Trust All to Desalination Plants," Blue Community Network News, March 24, 2025, online.
268. Wise Guy Reports, *Global Wastewater Market Research Report*, July 4, 2024, online.
269. Shilpa Tiku, "Market Outlook: Growth Opportunities in PFAS Water and Wastewater Treatment (2024–2031)," *Water Online*, February 24, 2025, online.
270. Juergen Voegele et al., "Scaling Up Finance to Ensure a Water-Secure Future for All," *World Bank Blog*, March 19, 2024, online.
271. "EPSU and AK Europa Joint Letter to EU Commission on the Water Resilience Strategy," EPSU, January 15, 2025, online.
272. *The Water Futures Market: Gambling with our Water* (Food & Water Watch December 2021), online.
273. Mariana Mazzucato et al., *The Economics of Water: Valuing the Hydrologic Cycle as a Global Common Good* (Global Commission on the Economics of Water, October 2024), online.

274. Claire Fyson, "The Dangers of Blue Carbon Offsets: From Hot Air to Hot Water?" *Climate Analytics*, November 16, 2017, online.
275. Erik Kristensen et al., "Predicting Climate Mitigation Through Carbon Burial in Blue Carbon Ecosystems — Challenges and Pitfalls," *Global Change Biology* 31, no. 1 (January 6, 2025), online.
276. "Blue Bond," European Commission, accessed September 2, 2025, online.
277. Jennifer L., "What Are Blue Carbon Credits? Everything You Need to Know," Carbon Credits, March 28, 2023, online.
278. Sara Kosmajac, "'First-Ever' Blue Bond in MENA Region Sees Light of Day," *Offshore Energy*, December 26, 2024, online.
279. OPR World, accessed September 2, 2025, online.
280. European Commission, "Commission Adopts Ocean Pact with €1 Billion to Protect Marine Life and Strengthen Blue Economy," press release, June 11, 2025, online.
281. Nathan Thanki & Serayna Solanki, "Marine Geoengineering: Between Profits and Climate Protection, Our Oceans Are Becoming an Experimental Field," *Geoengineering Monitor*, June 1, 2023, online.
282. Brianne Kelly, *Financing Coastal Blue Carbon in Canada: Potential Tools for Supporting Protection, Restoration and Stewardship of Blue Carbon Ecosystems* (World Wildlife Fund, 2022), online.
283. Brad Badelt, "Carbon Offsets Are Helping Protect B.C.'s Great Bear Rainforest. But Is That Sustainable?" *CBC*, January 7, 2024, online.
284. Zoe Brent et al., "The Illusion of 'Blue Growth,'" Transnational Institute, November 28, 2018, online.
285. Fyson, "The Dangers of Blue Carbon Offsets."
286. Dyna Rochmyaningsih, "Aquaculture Is Main Driver of Mangrove Losses," *Eco-Business*, June 28, 2017, online.
287. Zoe Brent et al., "The Illusion of 'Blue Growth.'"

288. Justin Villamil, "Ecuador's 'Blue Bond' Deal Won't Save the Galapagos," *Jacobin*, July 6, 2023, online.
289. Andre Standing, *Blue Finance: How Much Debt Can The Ocean Sustain? Implications for Coastal Fishing Communities in South Africa* (Transnational Institute, May 2023), online.
290. Raphaëlle Gauthier, "Drowning Out Voices: The Harmful Impact of Marine Protection Privatization on Indigenous Peoples," *EJIL: Talk!*, December 23, 2024, online.
291. Adam Smith, *The Wealth of Nations*, 2nd edition (Oxford, 1880).
292. Susan Nerberg, "I Am Mutehekau Shipu: A River's Journey to Personhood in Eastern Quebec," *Canadian Geographic*, April 8, 2022, online.
293. Phoebe Weston, "Jeff Bezos Fund Ends Support for Climate Group Amid Fears Billionaires 'Bowing Down' to Trump," *The Guardian*, February 6, 2026, online.
294. Oliver Milman, "US Banks Predict Climate Goals Will Fail — but Air Conditioning Firms Will Thrive," *The Guardian*, April 2, 2025, online.
295. UN News, "World Is 'On Notice' as Major UN Report Shows One Million Species Face Extinction," May 6, 2019, online.
296. Food and Agriculture Organization of the United Nations, accessed September 15, 2025, online.
297. George Monbiot, "We Need to Talk About Water — and the Fact That the World Is Running Out of It," *The Guardian*, March 4, 2024, online.
298. "News Spotlight: Fossil Fuels Behind Extreme Ocean Temperatures," *Conservation International*, January 30, 2025, online.
299. World Meteorological Organization, "WMO Report Documents Spiralling Weather and Climate Impacts," press release, March 19, 2025, online.
300. Sebastien Malo, "Air Pollution Is The World's Top Killer, According to New Research," Reuters, November 21, 2018.

301. Daisy Dunne, "Analysis: 95% of Countries Miss UN Deadline to Submit 2035 Climate Pledges," *Carbon Brief*, February 10, 2025, online.
302. Damian Carrington, "Climate Crisis on Track to Destroy Capitalism, Warns Top Insurer," *The Guardian*, April 3, 2025, online.
303. Oxfam International, *Takers Not Makers: The Unjust Poverty and Unearned Wealth of Colonialism*, 2025, online.
304. Jake Traylor and David Ingram, "Elon Musk, Jeff Bezos and Mark Zuckerberg to Attend Trump's Inauguration," *NBC News*, January 14, 2025, online.
305. Elon Musk (@elonmusk), "Without me, Trump would have lost the election, Dems would control the House and the Republicans would be 51-49 in the Senate," X, June 5, 2025, online.
306. Oxfam International, *Takers Not Makers*.
307. Save the Children, "2024 in Review: 35 Children Born into Hunger Every Minute in 2024," press release, December 30, 2024, online.
308. Oxfam International, "World's Poorest Countries to Slash Public Spending by More Than $220 Billion in Face of Crushing Debt," press release, October 9, 2023, online.
309. Keph Senett, "The Wealth Gap in Canada Is Wider Than Ever," *MoneySense*, July 24, 2025, online.
310. "Richest 1% Own 58% of Total Wealth in India: Oxfam," *The Economic Times*, January 16, 2017, online.
311. Antony Sguazzin, "South Africa Wealth Gap Unchanged Since Apartheid, Says World Inequality Lab," *Time*, August 5, 2021, online.
312. Sarah Schöngart et al., "High-Income Groups Disproportionately Contribute to Climate Extremes Worldwide," *Nature Climate Change* 15 (2025), online.
313. Buller, *The Value of a Whale*.

314. Food & Water Watch, "Cap and Trade: More Pollution for the Poor and People of Color," November 2019, online.
315. Ian Tucker, "Climatologist Friederike Otto: 'The More Unequal the Society Is, the More Severe the Climate Disaster,'" *The Guardian*, April 19, 2025, online.
316. Greenpeace International, *Bankrolling Ecosystem Destruction*, Greenpeace, March 26, 2024, online.
317. "Don't Be Fooled! Biodiversity Markets Are False Solutions," civil society statement on biodiversity offsets and credits, March 31, 2025, online.
318. Buller, *The Value of a Whale*.
319. Oxfam, *Takers Not Makers*.
320. C.J. Polychroniou, "The Global South Is on the Brink of a Disastrous Debt Crisis, Reform Is Urgent," *Truthout*, December 15, 2024, online.
321. Peoples of the Global South, *Manifesto for an Ecosocial Energy Transition from the Peoples of the South*, February 9, 2023, online.
322. Tax Justice Network, "Ministers, Economists Urge Honesty About Benefits of UN Tax Convention for Europe," press release, March 15, 2024, online.
323. Paul Krugman, "Tax the Rich and Help America's Children," *New York Times*, October 25, 2021, online.
324. Sam Pizzigati, "A Welcome New Twist on America's Tax Future," *Inequality.org*, April 12, 2025, online.
325. Mark Bou Mansour, "Countries Can Raise $2 Trillion by Copying Spain's Wealth Tax, Study Finds," *Tax Justice Network*, August 19, 2024, online.
326. Nam Pham et al., *Nature-Based Solutions* (Indigenous Environmental Network, November 2022), online.
327. Michael Igoe, "Can 'Nature-Based Solutions' Be More Than a Buzzword?" *Devex*, December 13, 2019, online.
328. Chris Vlavianos, "How Indigenous Peoples Are Safeguarding Earth's Biodiversity," Greenpeace, August 8, 2024, online.

329. Nemonte Nenquimo and Mitch Anderson, "Returning the Amazon Rainforest to Its True Caretakers," *Wired*, November 29, 2024, online.
330. "Canada's Boreal Forest: Why It's So Important," Natural Resources Defense Council, July 1, 2022, online.
331. "Forest Carbon," Government of Canada, accessed September 25, 2025, online.
332. Parks Canada, "Nunatsiavut Government, Makivvik, and Government of Canada Advance to Next Step Toward Establishing an Inuit Protected Area in the Waters of Northern Labrador," press release, November 29, 2024, online.
333. Canadian Press, "Canada Recognizes Aboriginal Title Over Haida Gwaii in Historic Agreement," *CBC*, February 17, 2025, online.
334. Cormac Cullinan, *Wild Law: A Manifesto for Earth Justice* (Green Books, 2003).
335. David R. Boyd, *The Rights of Nature: A Legal Revolution That Could Save the World* (ECW Press, 2017).
336. Katie Surma, "What Are the Rights of Nature?" *Inside Climate News*, April 2, 2025, online.
337. Eco Jurisprudence Monitor, accessed September 25, 2025, online.
338. Charlotte Graham-McLay, "A New Zealand Mountain Has Been Granted Personhood. Here's Why It Matters," Associated Press, January 30, 2025, online.
339. Assembly Bill A5156A, New York Senate, online.
340. Julien Greene, "Should The Yukon River Be Considered a 'Person' with Rights? Some Say It's an Idea Whose Time Has Come," *CBC*, May 3, 2025, online.
341. Robert Macfarlane, "Is This River Alive?," *The Guardian*, April 26, 2025, online.
342. Katie Surma, "Antarctica's Fate Will Impact the World. Is It Time to Give the Region a Voice at Climate Talks?" *Inside Climate News*, October 31, 2024, online.

343. Leslie-Anne Duvic-Paoli, *Biodiversity, Conservation and Human Rights: A Review* (German Institute for Human Rights, January 1, 2025), online.

344. *Implementing a Human Rights-Based Approach* (Forest Peoples Programme Biodiversity Working Group, March 2022), online.

345. David R. Boyd and Stephanie Keene, *Human Rights-Based Approaches to Conserving Biodiversity: Equitable, Effective and Imperative* (UNHR, August 2021), online.

346. Ashoka Mukpo, "Advocates Call for a New Human Rights-Based Approach to Conservation," *Mongabay*, October 12, 2021, online.

347. *Kunming-Montreal Global Biodiversity Framework* (Conference of the Parties to the Convention on Biological Diversity, December 2022), online.

348. Duvic-Paoli, *Biodiversity, Conservation and Human Rights.*

349. *Core Human Rights Principles for Private Conservation Organizations and Funders* (United Nations Environment Programme, December 9, 2024), online.

350. Annie Reneau, "'Forgotten' Water Harvesting System Transforms 'Barren Wasteland' into Thriving Farmland," *Upworthy*, March 6, 2025, online.

351. Agence France-Presse, "Huge River Restoration Effort Launched at UN Water Summit," March 23, 2023, online.

352. Phoebe Weston, "From Sewage and Scum to Swimming in 'Blue Gold': How Switzerland Transformed Its Rivers," *The Guardian*, March 17, 2025, online.

353. Benjamin Shingler, "'Sponge Cities': An Absorbing Idea in the Face of Climate Change," *CBC*, November 4, 2024, online.

354. *Removing Dams, Restoring Rivers* (World Wildlife Fund, February 28, 2023), online.

355. Lucy Sherriff, "'Anything That Can Be Built Can Be Taken Down': The Largest Dam Removal in US History Is Complete — What Happens Next?" *BBC*, September 3, 2024, online.

356. "We Cannot Trust All to Desalination Plants."
357. We Own It, accessed September 15, 2025, online.
358. *The Water Justice Manifesto* (The Peoples Water Forum, March 23, 2023), online.
359. António Guterres, "A Moment of Truth," speech at the American Museum of Natural History, June 5, 2024, online.
360. Doug Specht, "Only 15 Countries Have Met Their Latest Paris Agreement Deadline. Is Any Nation Serious About Tackling Climate Change?" *The Conversation*, March 19, 2025, online.
361. Kiley Bense et al., "In His First 100 Days, Trump Launched an 'All-Out Assault' on the Environment," *Inside Climate News*, April 30, 2025, online.
362. Guterres, "A Moment of Truth."
363. Tzeporah Berman, "Fossil Fuel Non-Proliferation Treaty Initiative: 'Never Underestimate the Power That We Have to Make Change,'" IKEA Foundation, May 24, 2024, online.
364. *Smoke and Mirrors: The Reality of Plastics Credits and Offsetting* (Break Free from Plastic, 2023), online.
365. "The Rarely Told Story of the Widely Used Water Sachets," UN Environment Programme, May 24, 2023, online.
366. "High Seas Treaty Ratification Tracker," accessed September 25, 2025, online.
367. "RE: Canada Should Be a World Leader in Protecting the High Seas," November 7, 2023, online.
368. Office of the United Nations Human Rights Commissioner, "Ocean Issues Are Human Rights Issues, Says UN Expert," press release, March 14, 2025, online.
369. "In Historic Move, UN Declares Healthy Environment a Human Right," UN Environmental Programme, news release, July 26, 2022, online.
370. "Transnational Corporate Accountability and Human Rights Treaty," Center for Environmental Law, January 2018, online.

371. Victoria Schneider, "World Bank Launches Historic Framework Addressing Harms from Development Projects," *Mongabay*, May 9, 2025, online.
372. Philip Drost, "Why These Island Nations Want Ecocide to Become an International Crime," *CBC*, March 22, 2025, online.

INDEX

Entertainment. Writing. Culture.

ECW is a proudly independent, Canadian-owned book publisher. We know great writing can improve people's lives, and we're passionate about sharing original, exciting, and insightful writing across genres.

Thanks for reading along!

We want our books not just to sustain our imaginations, but to help construct a healthier, more just world, and so we've become a certified B Corporation, meaning we meet a high standard of social and environmental responsibility — and we're going to keep aiming higher. We believe books can drive change, but the way we make them can too.

Being a B Corp means that the act of publishing this book should be a force for good — for the planet, for our communities, and for the people that worked to make this book. For example, everyone who worked on this book was paid at least a living wage. You can learn more at the Ontario Living Wage Network.

This book is also available as a Global Certified Accessible™ (GCA) ebook. ECW Press's ebooks are screen reader friendly and are built to meet the needs of those who are unable to read standard print due to blindness, low vision, dyslexia, or a physical disability.

This book is printed on FSC®-certified paper. It contains recycled materials, and other controlled sources, is processed chlorine free, and is manufactured using biogas energy.

FSC
www.fsc.org
MIX
Paper | Supporting responsible forestry
FSC® C103567

ECW's office is situated on land that was the traditional territory of many nations, including the Wendat, the Anishinaabeg, Haudenosaunee, Chippewa, Métis, and current treaty holders the Mississaugas of the Credit. In the 1880s, the land was developed as part of a growing community around St. Matthew's Anglican and other churches. Starting in the 1950s, our neighbourhood was transformed by immigrants fleeing the Vietnam War and Chinese Canadians dispossessed by the building of Nathan Phillips Square and the subsequent rise in real estate value in other Chinatowns. We are grateful to those who cared for the land before us and are proud to be working amidst this mix of cultures.

ecwpress.com